McGraw-Hill's

Math

GRADE 4

New York Chicago San Francisco Lisbon London Madrid Mexico City
Milan New Delhi San Juan Seoul Singapore Sydney Toronto

The McGraw·Hill Companies

1 2 3 4 5 6 7 8 9 10 11 12 13 14 15 DOW/DOW 1 9 8 7 6 5 4 3 2

ISBN   978-0-07-177560-1
MHID   0-07-177560-9

e-ISBN   978-0-07-177561-8
e-MHID   0-07-177561-7

**Cataloging-in-Publication data for this title are on file at the Library of
Congress.**

Printed and bound by RR Donnelley.

Editorial Services: Pencil Cup Press
Production Services: Jouve
Illustrator: Eileen Hine
Designer: Ella Hanna

McGraw-Hill books are available at special quantity discounts to use as
premiums and sales promotions or for use in corporate training programs. To
contact a representative, please e-mail us at bulksales@mcgraw-hill.com.

This book is printed on acid-free paper.

Table of Contents

Table of Contents

Welcome to McGraw-Hill's Math!

This book will help you succeed in your mathematics studies. Its short lessons explain key points and provide practice exercises.

Begin with the Pretest. This will help you determine whether you need to work on some skills more than others. Then read the Table of Contents. Seeing how a book is organized will help guide your work.

Look at the 10-Week Summer Study Plan. You can use it to plan your time. Remember, the Summer Study Plan is only a guide. Work at your own pace.

Each chapter ends with a Chapter Test. These tests will show you what you have mastered as well as the skills that you may need to practice more.

Finish your work with the Posttest. This will show you how well you have completed the program.

Remember to practice math. Practice will help you master the skills and will certainly make learning easier.

10-Week Summer Study Plan

Many students will use this book as a summer study program.
Use this 10-week study plan to help you plan your time.
Put a ✔ in the box when you have finished the day's work.

	Day	Lesson Pages	Test Pages
Week 1	Monday		Pretest 8–13
	Tuesday	14, 15, 16, 17	
	Wednesday	18, 19	20–21
	Thursday	22, 23	
	Friday	24, 25	
Week 2	Monday	26, 27	
	Tuesday	28, 29	30–31
	Wednesday	32, 33	
	Thursday	34, 35	
	Friday	36, 37	
Week 3	Monday	38, 39	
	Tuesday	40, 41	
	Wednesday	42	43–44
	Thursday	45, 46, 47	
	Friday	48, 49, 50	
Week 4	Monday	51	52–53
	Tuesday	54, 55, 56	
	Wednesday	57, 58, 59	
	Thursday	60	61–62
	Friday	63, 64, 65	
Week 5	Monday	66, 67, 68	
	Tuesday	69	70–71
	Wednesday	72, 73, 74	
	Thursday	75, 76	
	Friday	77, 78	79–80

	Day	Lesson Pages	Test Pages
Week 6	Monday	81, 82, 83	
	Tuesday	84, 85, 86	
	Wednesday	87, 88, 89	
	Thursday	90, 91	92–93
	Friday	94, 95, 96	
Week 7	Monday	97, 98, 99	
	Tuesday	100, 101	
	Wednesday	102	103–104
	Thursday	105, 106, 107	
	Friday	108, 109, 110	
Week 8	Monday	111, 112	
	Tuesday	113	114–115
	Wednesday	116, 117	
	Thursday	118, 119	
	Friday	120, 121	
Week 9	Monday	122, 123	124–125
	Tuesday	126, 127	
	Wednesday	128, 129	
	Thursday	130, 131	
	Friday	132, 133	
Week 10	Monday	134	135–136
	Tuesday	137, 138, 139	
	Wednesday	140, 141	
	Thursday	142	143–144
	Friday		Posttest 145–150

Name _____

Fill in each blank. Use the best term from the word bank. You may not use every term.

Word Bank

• even	• multiples	• odd
• round	• factors	• number line
• quotient	• sum	• divisor
• difference	• skip count	• product
• add	• subtract	• array
• equal groups	• multiply	• divide
• circle	• square	• triangle
• mental math	• fraction	• expression
• denominator	• numerator	• equation

1 A(n) _____ is a line that shows numbers in order.

2 The number 13 is a(n) _____ number.

3 In the number sentence 4 × 15 = 60, 60 is the _____.

4 You can _____ a number to estimate to the nearest 10 or 100.

5 When you count 5–10–15–20, you _____.

6 _____ are multiplied to find a product.

7 To find _____ of 5, you multiply numbers by 5.

8 The number 4 is a(n) _____ number.

9 An arrangement of objects into columns and rows is a(n) _____.

10 When you divide, the answer is called the _____.

11 The answer to a subtraction problem is called the _____.

12 A(n) _____ is a polygon with three sides.

13 A number sentence that shows two equal expressions and an equal sign is called a(n) _____.

14 The _____ is the number above the fraction bar in a fraction.

Write each number in standard form.

15 2 tens 8 ones

_____

16 thirty-two

_____

17 400 + 10 + 6

_____

18 fourteen

_____

19 3 hundreds 5 tens

_____

20 fifty-seven

_____

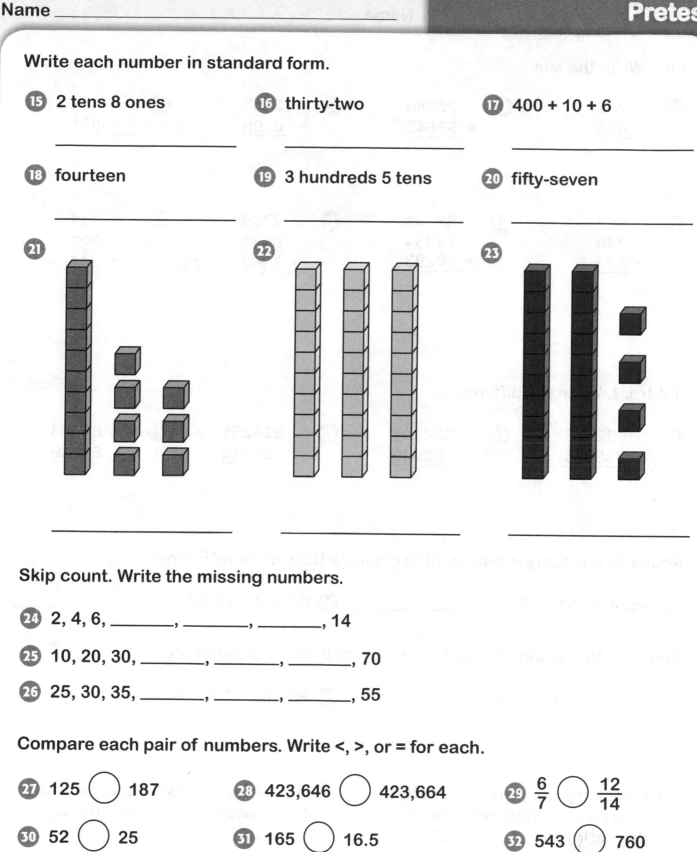

21

_____

22

_____

23

_____

Skip count. Write the missing numbers.

24 2, 4, 6, _____, _____, _____, 14

25 10, 20, 30, _____, _____, _____, 70

26 25, 30, 35, _____, _____, _____, 55

Compare each pair of numbers. Write <, >, or = for each.

27 125 ◯ 187

28 423,646 ◯ 423,664

29 $\frac{6}{7}$ ◯ $\frac{12}{14}$

30 52 ◯ 25

31 165 ◯ 16.5

32 543 ◯ 760

Name _____

Add. Write the sum.

33   745
       + 925

34   90289
       + 53648

35   15463
       + 8109

36   50901
       + 88521

37   913
         745
       + 823

38   66743
         78934
       + 60293

39   7769
         9091
       + 7362

40   534
         901
       + 441

Subtract. Write the difference.

41   932456
       − 738575

42   332103
       − 98943

43   934251
       − 83746

44   86754
       − 86539

Round to the nearest thousand to estimate the sum or difference.

45 45,567 − 31,339 = _____

46 93,854 + 59,201 = _____

Round to the nearest hundred to estimate the sum or difference.

47 56,710 + 732,820 = _____

48 89,560 − 7,431 = _____

Solve.

49 A rectangular yard is 17 meters long and 9 meters wide. What is the perimeter of the yard?

_____

50 A pool measures 11 feet long and 9 feet wide. What is the area of the pool?

_____

Find the product.

51 12 × 0 = _____   **52** 4 × 12 = _____   **53** 7 × 9 = _____

54 13 × 3 = _____   **55** 10 × 7 = _____   **56** 23 × 1 = _____

Find the quotient.

57 124 ÷ 3 = _____   **58** 45 ÷ 9 = _____   **59** 49 ÷ 7 = _____

60 144 ÷ 12 = _____   **61** 77 ÷ 7 = _____   **62** 110 ÷ 10 = _____

63 Which multiplication property is shown in this sentence?
17 × (2 + 4) = (17 × 2) + (17 × 4)

_____

Identify each number as prime or composite.

64 37 _____   **66** 35 _____

65 51 _____   **67** 22 _____

Find all the factor pairs of each number.

68 18 _____

69 23 _____

Solve. Write the expression you used to find the answer.

70 Ms. Williams fills a small fishbowl with 6 cups of water. She has a larger aquarium that holds 7 times as much water. How many cups of water will it take to fill the aquarium?

_____

71 Martha has 12 times as many blue marbles as red marbles. If she has 96 blue marbles, how many red marbles does she have? Use the variable y for the number of red marbles.

_____

Write an equivalent fraction for each.

72 $\dfrac{3}{5}$ _____

73 $\dfrac{40}{88}$ _____

74 $\dfrac{3}{51}$ _____

Add or subtract. Write your answers in simplest form.

75 $\dfrac{1}{7} + \dfrac{6}{7} =$ _____

76 $\dfrac{3}{12} + \dfrac{6}{12} =$ _____

77 $817\dfrac{11}{13} - 353\dfrac{4}{13} =$ _____

Look at each pattern. Draw the missing shape.

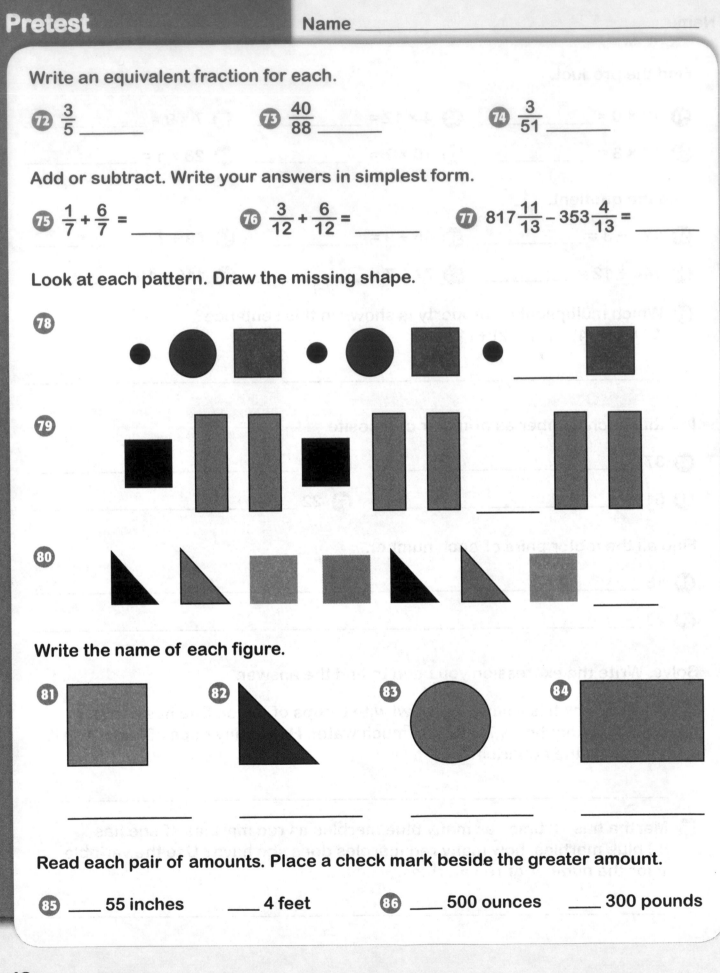

78

79

80

Write the name of each figure.

81   **82**   **83**   **84**

_____   _____   _____   _____

Read each pair of amounts. Place a check mark beside the greater amount.

85 ___ 55 inches   ___ 4 feet   **86** ___ 500 ounces   ___ 300 pounds

Solve and explain.

87 Is the sum of 67 + 29 equal to the sum of 29 + 67? Explain how you know.

_____

_____

88 Franz says that 4 × 5 = 20, so 4 × 500 = 200.

Is Franz correct? _____

Why or why not? _____

_____

_____

The graph shows the number of students practicing the guitar, the drums, the piano, and the trumpet at a small music school.

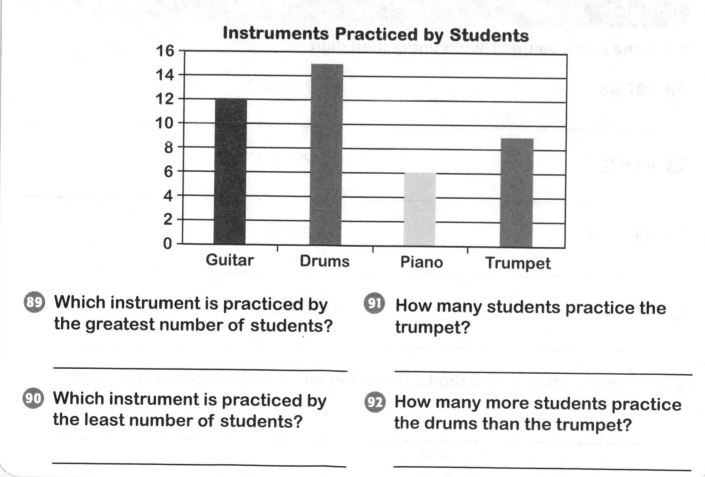

Instruments Practiced by Students

89 Which instrument is practiced by the greatest number of students?

_____

91 How many students practice the trumpet?

_____

90 Which instrument is practiced by the least number of students?

_____

92 How many more students practice the drums than the trumpet?

_____

Name _____

Place Value

Place value is the value of the place a digit holds in a number.

Example

The number 7,923 has four digits. The 7 is in the thousands place. The 9 is in the hundreds place. The 2 is in the tens place. The 3 is in the ones place.
So 7,923 means 7 thousands + 9 hundreds + 2 tens + 3 ones.

You can use a place-value chart to help you tell the value of a digit in a number.

hundred thousands	ten thousands	thousands	hundreds	tens	ones
4	9	3	2	8	6

This number is 493,286. The 2 is in the hundreds place. The 9 is in the ten thousands place.

Identify

Write the place value of each underlined digit.

1 207,8<u>3</u>1

_____hundreds_____

2 80,47<u>5</u>

_____

3 <u>9</u>30,134

_____

4 <u>1</u>3

_____

5 7<u>1</u>0,352

_____

6 2<u>9</u>,473

_____

7 2,<u>3</u>09

_____

8 6<u>0</u>3,914

_____

9 Write a number with 6 digits. Use a 4 in the thousands place and a 9 in the tens place.

_____

Writing Numbers

You can write numbers in different ways. You can use standard form, expanded form, and word form.

Example

You can write the standard form, expanded form, and word form for 439,287. Use a place-value chart to help.

Standard Form:
You write only the digits: 439,287

Expanded Form:
You write the value of each digit:
400,000 + 30,000 + 9,000 + 200 + 80 + 7

Word Form:
You write the number in word form:
four hundred thirty-nine thousand,
two hundred eighty-seven

millions	hundred thousands	ten thousands	thousands	hundreds	tens	ones
	4	3	9	2	8	7

thousands period	ones period

Write

Write each number in expanded form.

1 7,498

$\underline{7,000 + 400 + 90 + 8}$

2 five hundred seventy-one

_____

Write each number in standard form.

3 300,000 + 40,000 + 9,000 + 500 + 70 + 2

_____

4 seventy-two thousand, nine hundred sixty-three

_____

5 Write 500,000 + 30,000 + 1,000 + 700 + 90 + 4 in word form.

_____

6 What is the greatest five-digit number you can write? Write the number in standard form and expanded form.

_____

Place Value Relationships

The value of a digit changes when its place in a number changes. You can multiply or divide to determine the value.

Examples

In 700, the 7 has a value of 700. Move the 7 one place to the right What is the value of the digit now? You can use a place value chart to find the value.

hundreds	tens	ones
7	0	0

hundreds	tens	ones
	7	0

You can also divide by 10 to find the value: $700 \div 10 = 70$

The 7 has a value of 70.

In 942 the 9 has a value of 900. Move the 9 one place to the left. What is the value of the digit now? You can use a place value chart to find the value.

thousands	hundreds	tens	ones
	9	4	2

thousands	hundreds	tens	ones
9	0	4	2

You can also multiply by 10 to find the value: $900 \times 10 = 9{,}000$

The 9 has a value of 9,000.

Multiply and Divide

1 In 153, the 5 has a value of 50. Move the 5 one place to the left. What is the value of the digit 5 now?

_____500_____

2 In 782, the 7 has a value of 700. Move the 7 one place to the right. What is the value of the digit 7 now?

_____

3 In 680, the 6 has a value of 600. Move the 6 one place to the left. What is the value of the digit 6 now?

_____

4 In 658,942, the 4 has a value of 40. Move the 4 one place to the right. What is the value of the digit 4 now?

_____

5 In 47, the 4 has a value of 40. Move the 4 three places to the left. What is the value of the digit 4 now? How do you know?

_____

Comparing Numbers

You can compare numbers using place value.

Example

Which number is greater, 45,981 or 45,956?

Step 1: Write the numbers. Line up the digits. Start at the left and compare.

45,981
45,956

The ten thousands digit is the same in both numbers.

Step 2: Look at the next digit. Keep looking until you come to digits that are different.

45,981
45,956

The thousands digits and the hundreds digits are the same. The tens digits are different.

Step 3: Compare the digits that are different.

45,981
45,956

8 is greater than 5. So 45,981 is greater than 45,956.

The symbol > means greater than.
The symbol < means less than.

45,981 > 45,956

Compare

Compare each pair of numbers. Write <, >, or = for each.

1. 423 __<__ 856

2. 387 _____ 239

3. 8,235 _____ 4,901

4. 67,354 _____ 69,220

5. 712 _____ 2,364

6. 340,082 _____ 340,042

7. 53,713 _____ 53,713

8. 230,675 _____ 23,675

9. Write four numbers that are greater than 640,000 but less than 645,000.

_____

Name _____

Rounding Numbers

You can use place value to help you round numbers.

Examples

How can you round 37,923 to the nearest thousand?
Look at the digit to the right of the rounding digit.
If it is 5 or greater, add 1 to the rounding digit.
If it is less than 5, leave the digit alone.

37,9̲23

9 is greater than 5. Add 1 to the digit 7.
Change the digits to the right to zeros.

3̲8,000

37,923 rounded to the nearest thousand is 38,000.

How can you round 621,344 to the nearest hundred?
Look at the digit to the right of the rounding digit.
4 is less than 5.

621,3̲44

Leave the rounding digit alone.
Change the digits to the right to zeros.

621,3̲00

621,344 rounded to the nearest hundred is 621,300.

Round

Round each number to the place of the underlined digit.

1 83̲2 _____830_____

5 5̲4,233 _____

2 78̲4 _____

6 6̲50,174 _____

3 6,6̲88 _____

7 24̲8,535 _____

4 73,5̲42 _____

8 490,52̲6 _____

9 When rounded to the nearest thousand, which number would be rounded to 4,000? Place a check mark beside your answer.

_____ 3,481   _____ 3,542   _____ 5,000   _____ 4,769

Name _____

Problem Solving: Logical Reasoning

Sometimes, you can draw a chart to help organize the information to help you solve a problem.

Example

Grace, Dylan, Kira, and Diego are each wearing different colored shirts. Grace's shirt is red. Dylan's shirt is not white. Kira's shirt is not green. Diego's shirt is not yellow or white. What color shirt is each person wearing?

First, make a chart to show what you know.
- Each shirt is a different color.
- Grace's shirt is red.
- Dylan's shirt is not white.
- Kira's shirt is not green.
- Diego's shirt is not yellow or white.

	Red	White	Green	Yellow
Grace	yes	no	no	no
Dylan	no	no		
Kira	no		no	
Diego	no	no	yes	no

Then use reasoning and the information in the chart to complete the chart and find the answer.

Grace's shirt is red, so no other shirt can be red.
Diego's shirt is not red, white, or yellow, so it must be green.
Dylan's shirt must be yellow because it cannot be red, white, or green.
That means Kira's shirt must be white.

Solve

Draw a chart to help you solve the problem.

1. Ivan, Ester, Lloyd, and Cindy are each holding a marker that is either red, blue, black, or yellow. Ivan's marker is not red or black. Ester's marker is not red. Lloyd's marker is yellow. What color marker does each person have?

_____

_____

_____

Name _____

Write the place value of each underlined digit.

1 359,720

_____

4 601,243

_____

2 71,586

_____

5 8,382

_____

3 829,023

_____

6 4,521

_____

7 Write the number thirty-two thousand, seven hundred sixty-one in standard form.

_____

8 Write the same number in expanded form.

_____

9 Write the number 823,544 in expanded form.

_____

10 Write the same number in word form.

_____

_____

11 In 568,143, the 6 has a value of 60,000. Move the 6 one place to the left. What is the value of the digit 6 now?

_____

12 In 3,894, the 3 has a value of 3,000. Move the 3 one place to the right. What is the value of the digit 3 now?

_____

Compare each pair of numbers. Write <, >, or = for each.

⑬ 514 _____ 745

⑰ 1,989 _____ 3,241

⑭ 498 _____ 348

⑱ 531,973 _____ 531,973

⑮ 5,812 _____ 9,346

⑲ 875,231 _____ 875,213

⑯ 378,612 _____ 378,012

⑳ 916,222 _____ 916,217

Round each number to the place of the underlined digit.

㉑ 7̲21 _____

㉓ 5̲53,285 _____

㉒ 62,4̲31 _____

㉔ 3,5̲71 _____

Solve. Use another sheet of paper if you need to.

㉕ Bela, Jason, and Tom are playing a board game. Their scores are 20, 25, and 30 points. Jason does not have the most points. Bela does not have the fewest points. Tom has 30 points. How many points does each person have?

_____

㉖ Ben, Julia, Lin, and Ali are standing in line at school. Julia is first. Ali is not second or fourth. Lin is directly behind Ali. Ben is in front of Ali. In what order are the students in line?

_____

_____

_____

Name _____

Addition Properties

The properties of addition are the commutative, associative, and identity properties.

Commutative Property of Addition

You can add numbers in any order.
The sum remains the same.

$8 + 7 = 15$
$7 + 8 = 15$

Associative Property of Addition

You can change the grouping of addends.
The sum remains the same.

$(8 + 7) + 16 = 8 + (7 + 16)$
$15 + 16 = 8 + 23$
$31 = 31$

Remember! When adding, find the sum of numbers inside parentheses first.

Identity Property of Addition

Adding zero to a number does not change the number.

$16 + 0 = 16$

Complete and Identify

Complete each number sentence. Then identify the addition property shown in each number sentence.

1 $(8 + \underline{\quad 17 \quad}) + 23 = 8 + (17 + 23)$

Associative Property of Addition

2 $899 + \underline{\quad\quad} = 899$

_____

3 $45 + 31 = 31 + \underline{\quad\quad}$

_____

4 $14 + 6 = \underline{\quad\quad} + 14$

_____

5 $(24 + 12) + 8 = \underline{\quad\quad} + (12 + 8)$

_____

6 $(23 + 0) + 42 = \underline{\quad\quad} + 42$

_____

7 Pedro says that when zero is added to a number, the sum is always zero. Is he correct? Why or why not?

_____

_____

Relating Addition and Subtraction

Addition and subtraction have an inverse relationship. They can undo each other. Operations that have an inverse relationship are called inverse operations.

Example

Sasha has 12 baseballs. He finds 4 more at the field. How many baseballs does he have now?

$12 + 4 = 16$

Since addition and subtraction are inverse operations, you can use one to check the other. Subtract to check your answer.

$16 - 4 = 12$

The answer is the original number of baseballs. The answer is correct.

Add and Subtract

Add or subtract. Check your answer using the inverse operation.

1 $42 + 39 =$ _81_

  $81 - 39 = 42$

5 $38 - 12 =$ _____

_____

2 $15 + 10 =$ _____

_____

6 $29 - 4 =$ _____

_____

3 $8 + 23 =$ _____

_____

7 $40 - 9 =$ _____

_____

4 $32 + 4 =$ _____

_____

8 $23 - 11 =$ _____

_____

9 Lin read 66 pages of her book on Monday. How many more pages does she need to read to reach 85 pages?

_____

_____

Name _____

Adding Whole Numbers

You can add multidigit whole numbers. Sometimes you need to regroup.

Example

Add 23,674 + 6,870

Step 1: Add the ones. Regroup if necessary.

Step 2: Add the tens. Regroup if necessary.

Step 3: Add the hundreds, the thousands, the ten thousands, and so on. Regroup when you need to.

$$\begin{array}{r} 23674 \\ +\ 6870 \\ \hline 4 \end{array}$$

$$\begin{array}{r} 1 \\ 23674 \\ +\ 6870 \\ \hline 44 \end{array}$$

$$\begin{array}{r} 1\,1\,1 \\ 23674 \\ +\ 6870 \\ \hline 30544 \end{array}$$

The sum is 30,544.

Add

1.
$$\begin{array}{r} 491 \\ +\ 564 \\ \hline 1055 \end{array}$$

2.
$$\begin{array}{r} 4096 \\ +\ 8322 \\ \hline \end{array}$$

3.
$$\begin{array}{r} 52471 \\ +\ 29788 \\ \hline \end{array}$$

4.
$$\begin{array}{r} 70628 \\ +\ 4914 \\ \hline \end{array}$$

5.
$$\begin{array}{r} 504105 \\ +\ 81739 \\ \hline \end{array}$$

6.
$$\begin{array}{r} 623957 \\ +\ 348517 \\ \hline \end{array}$$

7. 38,231 + 25,877 = _____

9. 4,564 + 98,100 = _____

8. 620,873 + 76,455 = _____

10. 408,641 + 449,243 = _____

11. Look at problem 7 above. Is regrouping needed in the final step? Why or why not?

_____

_____

More Practice with Adding Whole Numbers

You know how to add two multidigit whole numbers. You can add more than two multidigit whole numbers the same way.

Example

Add 24,765 + 6,870 + 728 + 31,906

Step 1: Add the ones. Regroup if necessary.

```
    1
  24765
   6870
    728
+ 31906
      9
```

Step 2: Add the tens. Regroup if necessary.

```
   11
  24765
   6870
    728
+ 31906
     69
```

Step 3: Add the hundreds, the thousands, the ten thousands, and so on. Regroup if necessary.

```
  1311
  24765
   6870
    728
+ 31906
  64269
```

The sum is 64,269.

Add

1
```
    673
    822
+   742
   2237
```

2
```
   5824
   8237
+  6540
```

3
```
  413327
     826
+  70628
```

4
```
   61537
   35481
  322765
+   5825
```

5
```
   34693
   83001
   19548
+ 42965
```

6
```
  512868
    5906
  124850
+ 257623
```

7 Clearwater, Florida, had a population of 107,685 in 2010. Coral Springs had a population of 121,096. Orlando had a population of 238,300. What was the total population of the three communities?

_____

Subtracting Whole Numbers

You can subtract multidigit whole numbers. Sometimes you need to regroup.

Example

Find 63,003 – 25,392.

Step 1: Subtract the ones. Regroup if necessary.

```
  63003
- 25392
      1
```

Step 2: Subtract the tens, the hundreds, and so on. You may have to regroup more than once before you can subtract.

```
    12 9
  5 2 10 10
  63003
- 25392
  37611
```

Step 3: You can use addition to check your answer.

```
  1 1 1
  37611
+ 25392
  63003
```

The difference is 37,611.

Subtract

Use addition to check your answer.

1
```
  762
- 328
  434
```

2
```
  8295
- 3667
```

3
```
  87471
- 53743
```

4
```
  630119
- 228754
```

5
```
  926138
-  67880
```

6
```
  725914
- 699311
```

7 238,960 – 729 = _____

9 52,937 – 8,658 = _____

8 46,008 – 19,725 = _____

10 783,209 – 97,120 = _____

11 Look at problem 6. Why was the 0 in the hundred thousands place not written in the answer?

_____

Name _____

Addition and Subtraction Patterns

You can create patterns with repeated addition or subtraction. A chart can make these patterns easier to find.

Example

An expression contains numbers or variables and at least one operation.

A variable is a symbol that stands for a number.

$y + 5$ means "add 5."
$x - 4$ means "subtract 4."

variable
$$\overset{\text{variable}}{\underset{\text{expression}}{y + 5}}$$

You can use a chart to show expressions and repeated addition or subtraction. Use the expression to find the missing numbers.

$x - 4$
$5 - 4 = 1$

x	5	6	7	8
$x - 4$	1	2	3	4

Add or Subtract

Add or subtract to complete each table.

1

y	0	1	2	3
$y + 5$	5	6	7	8

2

x	0	12	24	36	48	60
$x + 12$						

3

a	30	27	24	21	18	15
$a - 3$						

4 Look at problem 2. What does $x + 12$ mean?

_____

Name _____

Additive Comparison

You can make comparisons using addition expressions and variables.

Example

Some new elephants arrive at a zoo. Louis the elephant is 23 years old. Betty the elephant is 29 years old. When Louis is 49 years old, how old will Betty be?

Step 1: How much older is Betty than Louis? You can write an expression to show this problem. Let y = how much older Betty is.

$$23 + y = 29$$

Step 2: Subtract to find the value of y.

$$29 - 23 = 6 \longrightarrow y = 6$$

Step 3: When Louis is 49 years old, how old will Betty be? Write an expression and use the value of y to find the answer.

$$49 + y \longrightarrow 49 + 6 = 55 \text{ years old}$$

Solve

Write the answer and the expressions you used to find it.

1 A young male elephant is 85 inches tall. A young female elephant is 57 inches tall. If the female elephant grows to 102 inches tall, how tall will the male elephant be if they grow the same amount?

_____

_____

2 A male box turtle is 17 years old. A female box turtle is 28 years old. When the male is 33 years old, how old will the female be?

_____

_____

Problem Solving: Rounding to Estimate

You can round numbers to estimate. An estimate is a careful guess.

Examples

Mr. Jones says there are 24,328 bees in a colony. Mr. Roy rounds the number to the nearest thousand and says there are about 24,000 bees in a colony. Mr. Roy's count is an estimate.

You can also round numbers to estimate sums and differences.

An arena has 19,763 open seats for hockey games. 10,433 people attend a game. About how many seats are empty?

You do not need an exact answer, because the question asks "**about** how many seats are empty." You can round to the nearest thousand and subtract.

19,763 rounded to the nearest thousand is 20,000.
10,433 rounded to the nearest thousand is 10,000.
20,000 – 10,000 = 10,000

About 10,000 seats are empty.

Remember! Your answer is an estimate. Use "about" in your answer.

Solve

Estimate to find the answer.

1. This week, 16,724 people watched a ball game at a stadium. Last week, 3,296 fewer people watched. About how many people watched a ball game at the stadium last week?

   about 14,000 people or 13,400 people
   _____

2. Joey was born in the year 2009. One of his older cousins was born in 1994. Rounding to the nearest ten, about how many years younger is Joey?

   _____

3. A high school club holds a bake sale to raise money for the band. 96 people came to the sale on Saturday. 145 people came to the sale on Sunday. About how many people came to the bake sale that weekend? Round to the nearest ten.

   _____

Name _____

Identify the property shown in each number sentence.

1 0 + 3,240 = 3,240

_____

4 (13 + 1) + 7 = 13 + (1 + 7)

_____

2 22,055 + 672 = 672 + 22,055

_____

5 (4 + 9) + 69 = 4 + (9 + 69)

_____

3 16 + (0 + 33) = 16 + 33

_____

6 18 + 40 = 40 + 18

_____

7 Janet tells her father that when zero is added to a number, the number does not change. She states that the rule is called the Commutative Property of Addition. Is she correct? Why or why not?

_____

_____

8 Which subtraction problem would you use to check the answer to 145 + 83 = 228? Place a check mark next to your answer.

_____ 145 − 83   _____ 228 − 83   _____ 83 + 228   _____ 83 + 145

9 Which addition problem would you use to check the answer to 598 − 266 = 332? Place a check mark next to your answer.

_____ 598 + 332   _____ 266 + 598   _____ 598 + 266   _____ 332 + 266

Add.

10
```
   734
+ 453
```

11
```
  61456
+ 30547
```

12
```
  53406
+  6732
```

13
```
  73375
  406909
+   7528
```

14
```
  441775
+ 410963
```

15
```
  8542
  3985
+ 4601
```

Name _____

Subtract.

16
$$\begin{array}{r} 873 \\ -\ 283 \\ \hline \end{array}$$

17
$$\begin{array}{r} 9528 \\ -\ 7336 \\ \hline \end{array}$$

18
$$\begin{array}{r} 48016 \\ -\ 25632 \\ \hline \end{array}$$

19
$$\begin{array}{r} 596744 \\ -\ 283475 \\ \hline \end{array}$$

20
$$\begin{array}{r} 745831 \\ -\ 56779 \\ \hline \end{array}$$

21
$$\begin{array}{r} 916550 \\ -\ 884062 \\ \hline \end{array}$$

Add or subtract to complete each table.

22

x	0	5	10	15	20	25
$x + 25$	25					

23

y	400	360	320	280	240
$y - 40$	360				

Solve. Show your work.

24 Gavin is 13 years old. His older cousin is 27 years old. When Gavin is 46 years old, how old will his cousin be?

_____

25 Megan spent 930 minutes last month swimming, and she spent 1,395 minutes doing homework. If she spends 1,240 minutes this month swimming, how many minutes will she spend doing homework if she increases the time spent by the same amount?

_____

_____

26 On Saturday, 11,794 people watched a rock concert at a stadium. On Sunday, 2,455 fewer people watched. About how many people watched a concert at the stadium on Sunday?

_____

Name _____

Multiplying by 2, 3, 4, and 5

When you know the number of equal groups and the number in each group, you can multiply to find the total. One way to think about multiplication is repeated addition. Another way is skip counting.

Example

Gary has some party hats. He arranges them in 3 rows. Each row has 4 hats. How many hats are there in all?

Repeated Addition: 4 + 4 + 4 = 12 hats

Skip Counting: 4, 8, 12 hats

Multiplication: 3 × 4 = 12 hats

Objects arranged in equal rows form an array.

The answer to a multiplication problem is called the product. The numbers that are multiplied are called factors.

factors $\left(8 \right) \times \left(3 \right) = \left(24 \right)$ product

Multiply

Find the product.

1 2 × 3 = ____6____

2 3 × 4 = _____

3 4 × 4 = _____

4 5 × 4 = _____

5 3 × 3 = _____

6 3 × 2 = _____

Write a multiplication sentence for each addition sentence.

7 3 + 3 + 3 = 9

8 5 + 5 + 5 + 5 = 20

9 2 + 2 = 4

_____

_____

_____

Multiplying by 6, 7, 8, and 9

You can multiply greater numbers too. You have to know the number of equal groups and the number in each group to find the product.

Examples

Zach is baking muffins. He has 8 rows with 3 muffins in each row. How many muffins does he have in all?

Repeated Addition:
3 + 3 + 3 + 3 + 3 + 3 + 3 + 3 = 24 muffins

Multiplication: 8 × 3 = 24 muffins

Mandy wants to serve the muffins. She puts them on 4 plates with 6 muffins on each plate. How many muffins does she have in all?

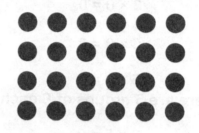

Repeated Addition:
6 + 6 + 6 + 6 = 24 muffins

Multiplication: 4 × 6 = 24 muffins

Multiply

Find the product.

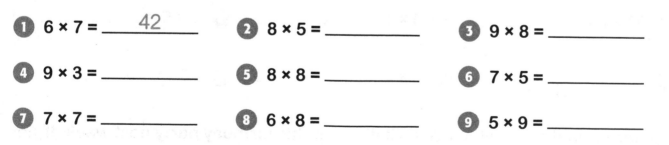

1 6 × 7 = ___42___

2 8 × 5 = _____

3 9 × 8 = _____

4 9 × 3 = _____

5 8 × 8 = _____

6 7 × 5 = _____

7 7 × 7 = _____

8 6 × 8 = _____

9 5 × 9 = _____

10 Hank arranges 8 rows of small pizzas on a large table. There are 7 pizzas in each row. How many pizzas are there in all?

_____

Name _____

Multiplying by 0 and 1

There is a simple pattern you can use to help you multiply when 0 or 1 are factors.

Examples

Multiplying by 0	**Multiplying by 1**
$1 \times 0 = 0$	$1 \times 1 = 1$
$2 \times 0 = 0$	$2 \times 1 = 2$
$3 \times 0 = 0$	$3 \times 1 = 3$
$4 \times 0 = 0$	$4 \times 1 = 4$
$5 \times 0 = 0$	$5 \times 1 = 5$

$5 \times 0 = 0$ $5 \times 1 = 5$

There are 5 groups of 0 each.   There are 5 groups of 1 each.

$0 \times 5 = 0$ $1 \times 5 = 5$

There are 0 groups of 5.   There is 1 group of 5.

Zero Property of Multiplication
Any number multiplied by 0 equals 0.

Identity Property of Multiplication
Any number multiplied by 1 equals that number.

Multiply

Find the product.

1 $0 \times 1 =$ _____0_____

2 $8 \times 1 =$ _____

3 $1 \times 9 =$ _____

4 $99 \times 0 =$ _____

5 $5 \times 1 =$ _____

6 $1 \times 5 =$ _____

7 $12 \times 1 =$ _____

8 $0 \times 45 =$ _____

9 $550 \times 1 =$ _____

10 Aaron wants to mail out 26 invitations for his birthday party next week. If he needs 1 stamp for each invitation, then how many stamps does Aaron need in all?

_____

Multiplying by 10, 11, and 12

Number patterns can help you multiply by 10, 11, and 12.

Examples

Multiplying by 10	**Multiplying by 11**	**Multiplying by 12**
10 × 1 = 10	11 × 1 = 11	12 × 1 = 12
10 × 2 = 20	11 × 2 = 22	12 × 2 = 24
10 × 3 = 30	11 × 3 = 33	12 × 3 = 36
10 × 4 = 40	11 × 4 = 44	12 × 4 = 48
10 × 5 = 50	11 × 5 = 55	12 × 5 = 60

Place a zero to the right of the number. The zero becomes a new ones digit.

Multiply the factor that is not 11 by 10. Add the factor to the product.

$11 × 3 = (10 × 3) + 3 = 33$

Break apart 12.
$12 = 10 + 2$

$12 × 4 = (10 × 4) + (2 × 4)$
$12 × 4 = 40 + 8$
$12 × 4 = 48$

Multiply

Find the product.

1 $10 × 7 =$ ___70___

2 $11 × 5 =$ _____

3 $12 × 8 =$ _____

4 $9 × 11 =$ _____

5 $10 × 8 =$ _____

6 $11 × 11 =$ _____

7 $2 × 10 =$ _____

8 $7 × 11 =$ _____

9 $12 × 3 =$ _____

10 $12 × 6 =$ _____

11 $10 × 3 =$ _____

12 $5 × 12 =$ _____

13 Explain how you can use $8 × 10$ to help solve $8 × 12$.

_____

_____

_____

Name _____

Identity and Commutative Properties of Multiplication

Multiplication properties can help you multiply.

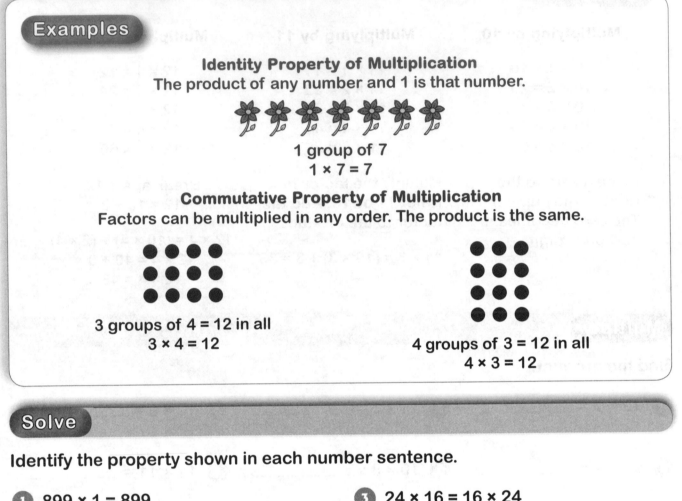

Examples

Identity Property of Multiplication
The product of any number and 1 is that number.

1 group of 7
$1 \times 7 = 7$

Commutative Property of Multiplication
Factors can be multiplied in any order. The product is the same.

3 groups of 4 = 12 in all
$3 \times 4 = 12$

4 groups of 3 = 12 in all
$4 \times 3 = 12$

Solve

Identify the property shown in each number sentence.

1 $899 \times 1 = 899$

Identity
_____

3 $24 \times 16 = 16 \times 24$

_____

2 $12 \times 49 = 49 \times 12$

_____

4 $2455 \times 1 = 2455$

_____

Solve.

5 Mr. Hunt's class has 28 students. He wants to put the desks in equal groups. The desks are already in 4 rows of 7 each. What is another way he can arrange the desks?

_____

Associative and Distributive Properties of Multiplication

There are other properties that can help you multiply.

Examples

Associative Property of Multiplication

You can multiply three numbers. When you group them in different ways, the product remains the same.

$(4 \times 2) \times 5$

$8 \times 5 = 40$

$4 \times (2 \times 5)$

$4 \times 10 = 40$

Distributive Property of Multiplication

You can multiply two addends by a factor. You can also multiply each addend by the same factor and add the products. The final total is the same.

$$(3 + 4) \times 2 = (3 \times 2) + (4 \times 2)$$

You can use the Distributive Property to break apart facts. Breaking apart facts can make multiplying easier.

Find 6×8.

$6 = 5 + 1$

So $(5 + 1) \times 8 = (5 \times 8) + (1 \times 8)$

$5 \times 8 = 40$

$1 \times 8 = 8$

$40 + 8 = 48$

$6 \times 8 = 48$

Solve

Identify the property used in each number sentence.

1 $(12 \times 2) \times 2 = 12 \times (2 \times 2)$

Associative
_____

3 $8 \times (9 \times 3) = (8 \times 9) \times 3$

_____

2 $(10 + 2) \times 6 = (10 \times 6) + (2 \times 6)$

_____

4 $7 \times (3 + 2) = (7 \times 3) + (7 \times 2)$

_____

Use the Distributive Property to break the fact into two easier facts.

5 $12 \times 5 =$

_____

6 $6 \times 9 =$

_____

Name _____

Dividing by 1, 2, 3, 4, and 5

You can divide to find a number of objects in each group or a number of equal groups.

Example

A group of 4 friends have 24 baseball cards. They want to share the cards equally with each other. How many cards should each friend get?

Think of sharing the cards equally among the friends.

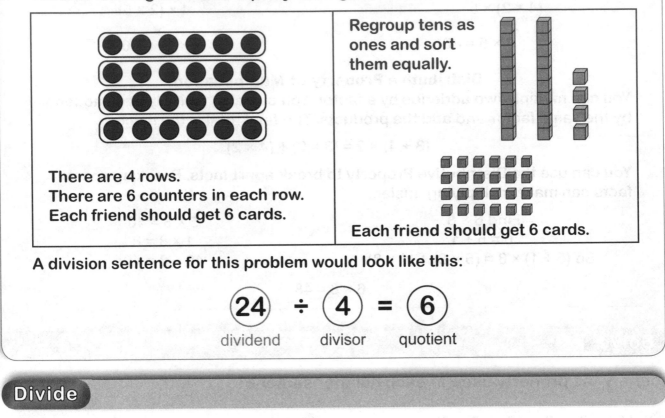

There are 4 rows.
There are 6 counters in each row.
Each friend should get 6 cards.

Regroup tens as ones and sort them equally.

Each friend should get 6 cards.

A division sentence for this problem would look like this:

$$24 \div 4 = 6$$

dividend   divisor   quotient

Divide

Find the quotient.

① 9 ÷ 3 = _____3_____   ② 28 ÷ 4 = _____   ③ 25 ÷ 5 = _____

④ 16 ÷ 4 = _____   ⑤ 12 ÷ 2 = _____   ⑥ 6 ÷ 1 = _____

Solve.

⑦ A deli has 12 rolls. The owner wants to show the rolls on 4 trays with an equal number of rolls on each tray. How many rolls should she put on

each tray? _____

Dividing by 6, 7, 8, and 9

Division helps find the number in each group or the number of equal groups.

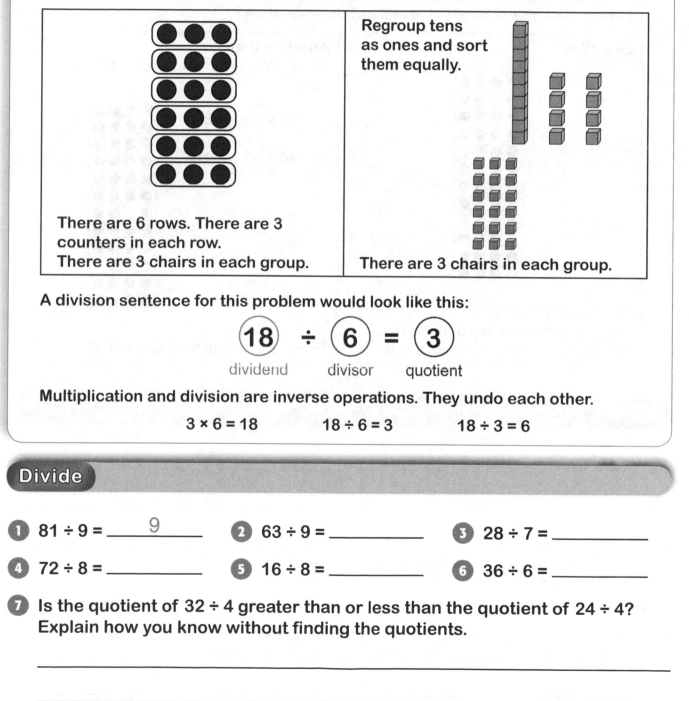

Example

Seth is arranging 18 chairs into 6 equal groups. How many chairs are in each group?

Regroup tens as ones and sort them equally.

There are 6 rows. There are 3 counters in each row.
There are 3 chairs in each group.

There are 3 chairs in each group.

A division sentence for this problem would look like this:

$$18 \div 6 = 3$$

dividend   divisor   quotient

Multiplication and division are inverse operations. They undo each other.

$3 \times 6 = 18$ $18 \div 6 = 3$ $18 \div 3 = 6$

Divide

1. $81 \div 9 = \underline{\quad 9 \quad}$

2. $63 \div 9 = \underline{\qquad}$

3. $28 \div 7 = \underline{\qquad}$

4. $72 \div 8 = \underline{\qquad}$

5. $16 \div 8 = \underline{\qquad}$

6. $36 \div 6 = \underline{\qquad}$

7. Is the quotient of $32 \div 4$ greater than or less than the quotient of $24 \div 4$? Explain how you know without finding the quotients.

_____

_____

Name _____

Dividing by 10, 11, and 12

Multiplication facts can help you solve division facts. Multiplication and division are inverse operations.

Example

Mr. Richard's classroom has 48 books. He equally divides the books onto 12 shelves. How many books does he put on each shelf?

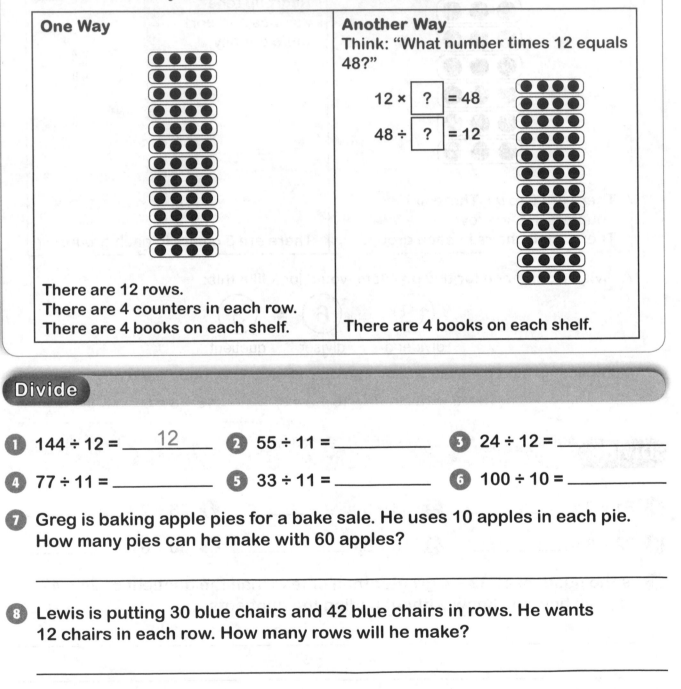

One Way

There are 12 rows.
There are 4 counters in each row.
There are 4 books on each shelf.

Another Way
Think: "What number times 12 equals 48?"

$$12 \times \boxed{?} = 48$$

$$48 \div \boxed{?} = 12$$

There are 4 books on each shelf.

Divide

1. $144 \div 12 = \underline{\quad 12 \quad}$

2. $55 \div 11 = \underline{\qquad}$

3. $24 \div 12 = \underline{\qquad}$

4. $77 \div 11 = \underline{\qquad}$

5. $33 \div 11 = \underline{\qquad}$

6. $100 \div 10 = \underline{\qquad}$

7. Greg is baking apple pies for a bake sale. He uses 10 apples in each pie. How many pies can he make with 60 apples?

_____

8. Lewis is putting 30 blue chairs and 42 blue chairs in rows. He wants 12 chairs in each row. How many rows will he make?

_____

Multiplication and Division Patterns

Sometimes multiplication and division facts create a pattern. A chart can be used in two ways to identify a pattern.

Examples

One way is showing how the numbers on one side of the chart are related to the numbers on the other side of the chart. This relationship is usually shown as an expression.

y	$y \times 4$
2	8
4	16
6	24

Another way is looking at the two sets of numbers and determining the expression that explains how the numbers are related.

y	$y \div 5$
15	3
30	6
45	9

Multiply or Divide

Use the rule to fill in the chart.

1

y	$y \times 9$
3	27
4	36
5	45

2

c	$c \times 12$
5	
6	
7	

3

x	$x \div 5$
50	
40	
30	

Identify the rule that explains the chart.

4

x	
120	12
100	10
80	8

5

y	
77	11
56	8
42	6

6

c	
3	6
6	12
9	18

Name _____

Problem Solving: Draw a Picture and Write an Equation

You can solve some problems by drawing pictures and writing equations. A picture can help you see a problem. It is a good way to check your answer.

Example

Read the Problem	Draw	Write an Equation
Mr. Henderson is making bouquets. He needs 6 flowers for each bouquet. Gina orders 5 bouquets. How many flowers will Mr. Henderson need to make the bouquets?	(drawing of flowers)	$6 \times 5 = 30$ flowers You can count the flowers in your drawing to check your answer.

Solve

Draw a picture and write an equation to solve each problem.

1. Jessica read 3 books every month. How many books did she read in 6 months?

   $3 \times 6 = 18$ books

2. Hector has 121 coins in his collection. He can fit 11 in a display case. How many display cases will he need to display his entire collection?

   _____

3. Jim is planting a garden with 5 seeds in each row. If he plants 40 seeds in all, how many rows are in Jim's garden?

   _____

Name _____

Multiply.

1 10 × 4 = _____

2 9 × 5 = _____

3 8 × 8 = _____

4 3 × 11 = _____

5 2 × 8 = _____

6 6 × 6 = _____

7 2 × 6 = _____

8 7 × 3 = _____

9 12 × 9 = _____

Divide.

10 48 ÷ 6 = _____

11 81 ÷ 9 = _____

12 120 ÷ 10 = _____

13 32 ÷ 8 = _____

14 9 ÷ 3 = _____

15 24 ÷ 12 = _____

16 63 ÷ 7 = _____

17 28 ÷ 4 = _____

18 25 ÷ 5 = _____

Solve. Show the equation you used to find the answer.

19 Anna spent $40 on movie tickets for herself and her friends. She bought 8 tickets in all. How much does 1 ticket cost?

_____

20 Mr. Park's fourth grade class is going on a field trip. He wants to put the 24 students in his class into 6 equal groups. How many students should he put in each group?

_____

21 Jamie wants to count her button collection. She makes 7 rows with 5 buttons in each row. How many buttons does she have in all?

_____

Name _____

22 Explain how knowing that 9 × 7 = 63 can help you solve 63 ÷ 7.

_____

_____

23 Mitchell writes 4 × 5 to describe the position of the desks in the classroom. Catie writes 5 × 4. Who is correct? Explain.

_____

_____

Identify the property shown in each number sentence.

24 129 × 1 = 129

_____

25 (8 + 4) × 5 = (8 × 5) + (4 × 5)

_____

26 (3 × 24) × 8 = 3 × (24 × 8)

_____

27 8 × 50 = 50 × 8

_____

28 102 × 25 = 25 × 102

_____

29 15 × 9 = (10 × 9) + (5 × 9)

_____

Use a multiplication property to solve.

30 (12 + 4) × 6 = _____

31 404 × 1 = _____

Use each rule to fill in the chart.

32

y	y × 5
2	
4	
6	
8	
10	

33

c	c × 2
2	
5	
8	
11	
12	

34

x	x ÷ 9
99	
81	
54	
27	
9	

Prime and Composite Numbers

Examples

A whole number greater than 1 that has exactly two factors—1 and itself— is a prime number.

3 and 5 are prime numbers.

$$3 = 1 \times 3$$
$$5 = 1 \times 5$$

There is only one way to make an array to show 3 or 5.

● ● ● ● ● 1×5

● ● ● 1×3

1 is neither prime nor composite.

A whole number greater than 1 that has more than two factors is a composite number.

6 is a composite number.

$$6 = 1 \times 6$$
$$6 = 2 \times 3$$

There is more than one way to make an array to show 6.

● ● ● ● ● ● 1×6

● ● ●
● ● ● 2×3

Identify

Identify each number as prime or composite.

1 4 ___composite___

2 7 _____

3 12 _____

4 29 _____

5 58 _____

6 23 _____

7 Taylor said that if you multiply two prime numbers, the product is also a prime number. Do you agree? Why or why not?

_____

_____

8 Kayla asked her friends what season they liked best. 15 friends said summer. 11 friends said winter. 8 friends said spring. 4 friends said fall. Which season did a prime number of friends vote for?

_____

Name _____

Factor Pairs

Numbers can have many factors. You can use multiplication to find all the factors of a number.

Example

Mr. Brown has 12 cans of soup. He wants to arrange them in equal-sized groups on a table. What are the ways he can arrange the cans? He needs to find all the factors of 12.

$12 = 1 \times 12$ or 12×1
He can arrange 12 groups of 1 can each or 1 group of 12 cans.

1 and 12 are factors of 12.

$12 = 2 \times 6$ or 6×2
He can arrange 2 groups of 6 cans each or 6 groups of 2 cans each.

2 and 6 are factors of 12.

$12 = 3 \times 4$ or 4×3
He can arrange 3 groups of 4 cans each or 4 groups of 3 cans each.

3 and 4 are factors of 12.

The factor pairs for 12 are: 1 and 12; 2 and 6; 3 and 4.

Factor

Find all the factor pairs of each number.

1. 7 ____  1 and 7 ____

4. 22 _____

2. 6 _____

5. 35 _____

3. 16 _____

6. 48 _____

7. Mr. Brown now has 100 cans. He wants to arrange them on a large table in his store. What are all the factor pairs of 100?

_____

_____

8. Antonio says that 2 is always a factor for even numbers. Is he correct? Why or why not?

_____

Multiples of 2, 5, and 9

The product of any two whole numbers is a multiple.

Examples

All multiples of 2 are even numbers.

$1 \times 2 = 2$
$2 \times 2 = 4$
$3 \times 2 = 6$
$4 \times 2 = 8$

All multiples of 5 have a 5 or 0 in the ones place.

$1 \times 5 = 5$
$2 \times 5 = 10$
$3 \times 5 = 15$
$4 \times 5 = 20$

The digits of all multiples of 9 add to 9 or a multiple of 9.

$1 \times 9 = 9$
$2 \times 9 = 18$
$3 \times 9 = 27$
$4 \times 9 = 36$

For 27, $2 + 7 = 9$, so 27 is a multiple of 9.

Identify

Write yes or no for each question.

1. Is 12 a multiple of 2? ____yes____

2. Is 40 a multiple of 5? _____

3. Is 76 a multiple of 5? _____

4. Is 30 a multiple of 2? _____

5. Is 43 a multiple of 9? _____

6. Is 54 a multiple of 9? _____

7. Is 82 a multiple of 5? _____

8. Is 75 a multiple of 2? _____

9. How do you know that 71 is not a multiple of 2? Explain.

_____

_____

_____

10. How do you know that 45 is a multiple of both 5 and 9? Explain.

_____

_____

_____

Making Comparisons Using Multiplication

You can make comparisons using multiplication expressions.

Example

Latoya has 6 coins. Katie has 4 times as many coins as Latoya. How many coins does Katie have?

Step 1: You can write an expression to show this problem.

$$6 \times 4$$

Step 2: Multiply.

$$6 \times 4 = 24$$

Katie has 24 coins.

Solve

Write the answer, and the expression used to find it.

1. Mr. Novak is making a large pot of soup. His recipe calls for 3 times as many beans as noodles. If he uses 2 cups of noodles, how many cups of beans will he need?

   $2 \times 3 = 6$ cups of beans

2. Rob and Ryan swim laps in a pool. Ryan swims 3 times as many laps as Rob. If Rob swims 7 laps, how many laps does Ryan swim?

   _____

3. Ms. Moreno has 4 times as many spoons as plates. If she has 5 plates, how many spoons does she have?

   _____

4. A store has 9 times as many shirts for sale as hats. If the store has 11 hats for sale, how many shirts are for sale?

   _____

Comparing Through Multiplication and Addition

You can solve some problems using multiplicative comparison or additive comparison. You have to read the problem carefully to know which to use. Multiplicative comparison questions often use the word "times" in the problem.

Examples

Additive Comparison

A male swan is 22 years old. A female swan is 31 years old. When the male is 59, how old will the female be?

$$22 + y = 31$$
$$31 - 22 = 9 \longrightarrow y = 9$$
$$59 + 9 = 68$$

The female will be 68 years old.

Multiplicative Comparison

Nicole has 6 grapes. Andy has 3 times as many grapes as Nicole. How many grapes does Andy have?

$$6 \times 3 = 18 \text{ grapes}$$

Andy has 18 grapes.

Solve

Write the answer, and the expression used to find it.

1 Ms. Link's recipe calls for 4 times as many peas as onions. If she uses 2 cups of onions, how many cups of peas will she need?

$4 \times 2 = 8$ cups of peas
_____

2 Ava builds a model of a racecar. The model is 1 foot long. The actual racecar is 13 times as long as the model. How long is the racecar?

_____

3 Luis biked 455 yards today, and Alex biked 298 yards. If Alex bikes 475 yards tomorrow, how far will Luis bike if they increase their length by the same amount?

_____

4 Haru has 9 marbles. Damon has 5 times as many marbles as Haru. How many marbles does Damon have?

_____

Name _____

Problem Solving: Variables

In some multiplicative comparison problems, you do not know both factors. You can use a variable to represent the missing factor. Then you can divide to find the value of the missing factor.

Example

John and his mom are working in a garden. His mom has 6 times as many seeds to plant as John. If she has 24 seeds, how many seeds does John have?

Step 1: You can write a number sentence to show this problem. Let $y =$ the number of John's seeds.

$$y \times 6 = 24$$

Step 2: Divide to find the value of y.

$$24 \div 6 = 4 \longrightarrow y = 4$$

John has 4 seeds.

Step 3: You can multiply to check your answer.

$$4 \times 6 = 24$$

The answer checks.

Solve

Write the answer and the expression used to find it.

1 Mr. Young's recipe calls for 6 times as much lettuce as peas. If he has 12 cups of lettuce, how many cups of peas does he need?

$y \times 6 = 12; 12 \div 6 = 2$ cups of peas

2 A carousel has 8 times as many red horses as blue horses. If there are 32 red horses on the carousel, how many blue horses are there?

_____

3 Meg has 36 oranges. She puts them into 4 bags. If she puts an equal number of oranges in each bag, how many oranges are in each bag?

_____

Problem Solving: Multistep Problems

Sometimes you have to answer one problem to solve another problem.

Example

Problem 1: On Sunday, Hugo and Ashley ran 4 miles in the morning and 3 miles in the afternoon. How many miles did they run in all?

Problem 2: They ran the same number of miles on Monday, Tuesday, and Wednesday. How far did they run during those three days?

Step 1: Read the first problem. Do you need to add, subtract, multiply, or divide? Write a number sentence and draw a picture to show the problem.

Step 2: Use the answer from the first problem to answer the second problem. Do you need to add, subtract, multiply, or divide? Write a number sentence and draw a picture.

```
        ?
 ┌───────────┐
 │  4  │  3  │
 └───────────┘
   4 + 3 = 7
```

Hugo and Ashley ran 7 miles on Sunday.

```
           ?
 ┌─────────────────┐
 │  7  │  7  │  7  │
 └─────────────────┘
      3 × 7 = 21
```

Hugo and Ashley ran a total of 21 miles on Monday, Tuesday, and Wednesday.

Solve

Write the answer and the number sentences you use.

1 Eric mows 4 lawns each week. Peter mows 3 times as many lawns as Eric. How many lawns does Peter mow each week?

$3 \times 4 = 12$ lawns

_____

Peter gets paid 8 dollars per lawn. How much does Peter earn each week?

_____

2 Mrs. Kumar puts one vase on each of 6 tables. There are 5 flowers in each vase. How many flowers does she use?

_____

19 of the flowers are yellow. The rest are white. How many flowers are white?

_____

Name _____

Identify each number as prime or composite.

1 16 _____ **4** 61 _____

2 13 _____ **5** 90 _____

3 8 _____ **6** 79 _____

Find all the factor pairs of each number.

7 7 _____

8 18 _____

9 63 _____

Write "yes" or "no" for each question.

10 Is 15 a multiple of 2? _____

11 Is 75 a multiple of 5? _____

12 Is 72 a multiple of 9? _____

13 Is 54 a multiple of 5? _____

Solve. Write the expressions you use and the answers.

14 Ms. Williams is making a large pot of chili for a party. The recipe calls for 5 times as many beans as peppers. If she uses 2 cups of peppers, how many cups of beans will she need?

_____

15 Jada made a model of an alligator for a science report. The model was 3 feet long. The actual alligator is 4 times as long as the model. How long is the alligator?

_____

Solve.

16 Jason has 8 markers. Kaya has 7 times as many markers as Jason. How many markers does Kaya have?

_____

17 Madison is 13 years old. Her brother is 19 years old. When Madison is 38 years old, how old will her brother be?

_____

18 Henry took 9 pictures in the morning and his mom took 22. If Henry takes 17 pictures in the afternoon, how many pictures would his mom take if the amount of pictures they take increases by the same amount?

_____

19 A zoo has 6 times as many horses as lions. If there are 36 horses at the zoo, how many lions are there?

_____

20 Jasmine and Andy walk dogs to earn money. Jasmine walks 3 dogs a month. Andy walks 4 times as many dogs as Jasmine. How many dogs does Andy walk in a month?

_____

Andy gets paid $9 for each dog he walks. How much money does Andy earn each month?

_____

21 Mr. Hollins has 7 red books and 11 blue books. How many books does he have in all?

_____

He wants to put the books on 3 shelves. He wants an equal number of books on each shelf. How many books should he place on each shelf?

_____

Multiplying a Two-Digit Number

You can multiply a two-digit number by a one-digit number. Drawing pictures can help you.

Example

Ms. Perez buys 3 bags of apples. There are 15 apples in each bag. How many apples does Ms. Perez buy?

Draw a picture to show 3 × 15.

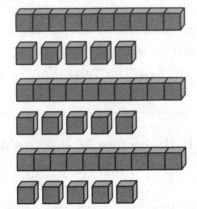

$3 \times 10 = 30$ $3 \times 5 = 15$

$30 + 15 = 45$

30 and 15 are partial products.

Ms. Perez buys 45 apples.

You can record multiplication using partial products. Multiply the ones. Write the product. Then multiply the tens. Write the product. Add the two partial products.

$$\begin{array}{r} 10 \\ \times\ 3 \\ \hline 30 \end{array} \quad + \quad \begin{array}{r} 5 \\ \times\ 3 \\ \hline 15 \end{array} \quad \longrightarrow \quad \begin{array}{r} 30 \\ +\ 15 \\ \hline 45 \end{array}$$

Multiply

Find each product using partial products.

1
$$\begin{array}{r} 13 \\ \times\ 4 \\ \hline 52 \end{array}$$

2
$$\begin{array}{r} 19 \\ \times\ 5 \\ \hline \end{array}$$

3
$$\begin{array}{r} 22 \\ \times\ 6 \\ \hline \end{array}$$

4
$$\begin{array}{r} 90 \\ \times\ 6 \\ \hline \end{array}$$

5 The owner of a restaurant buys 26 tables. She wants to buy 5 chairs for each table. How many chairs should she buy in all?

_____

Multiplying a Three-Digit Number

You can multiply a three-digit number by a one-digit number.

Example

David has 4 jars of coins. Each jar has 287 coins. How many coins does David have in all?

Step 1:
Multiply the ones. Regroup if you need to.

$$\begin{array}{r} 2 \\ 287 \\ \times 4 \\ \hline 8 \end{array}$$

4 × 7 ones =
2 tens and 8 ones

Step 2:
Multiply the tens. Be sure to add any extra tens. Regroup if you need to.

$$\begin{array}{r} 3\,2 \\ 287 \\ \times 4 \\ \hline 48 \end{array}$$

(4 × 8 tens) + 2 tens =
3 hundreds and 4 tens

Step 3:
Multiply the hundreds. Be sure to add any extra hundreds.

$$\begin{array}{r} 3\,1 \\ 287 \\ \times 4 \\ \hline 1148 \end{array}$$

(4 × 2 hundreds) +
3 hundreds =
1 thousand and
1 hundred

4 × 287 = 1148   David has 1,148 coins in all.

Multiply

Find each product.

1
$$\begin{array}{r} 142 \\ \times 5 \\ \hline 710 \end{array}$$

2
$$\begin{array}{r} 271 \\ \times 3 \\ \hline \end{array}$$

3
$$\begin{array}{r} 309 \\ \times 8 \\ \hline \end{array}$$

4
$$\begin{array}{r} 294 \\ \times 7 \\ \hline \end{array}$$

5
$$\begin{array}{r} 516 \\ \times 4 \\ \hline \end{array}$$

6
$$\begin{array}{r} 878 \\ \times 9 \\ \hline \end{array}$$

7 A giant panda weighs 236 pounds. An Arabian camel weighs 6 times as much as the panda. How much does the Arabian camel weigh?

_____

Name _____

Multiplying a Four-Digit Number

You can multiply a four-digit number by a one-digit number.

Example

A truck driver travels 2,182 miles each week. If he drives for 5 weeks, how many miles does he travel in all?

Steps 1 and 2:
Multiply the ones and tens. Regroup if you need to.

$$
\begin{array}{r}
4\,1 \\
2182 \\
\times \quad 5 \\
\hline
10
\end{array}
$$

Step 3:
Multiply the hundreds. Regroup if you need to.

$$
\begin{array}{r}
4\,1 \\
2182 \\
\times \quad 5 \\
\hline
910
\end{array}
$$

Step 4:
Multiply the thousands.

$$
\begin{array}{r}
4\,1 \\
2182 \\
\times \quad 5 \\
\hline
10910
\end{array}
$$

The truck driver travels 10,910 miles.

Multiply

Find each product.

1
$$
\begin{array}{r}
1290 \\
\times \quad 4 \\
\hline
5160
\end{array}
$$

2
$$
\begin{array}{r}
3403 \\
\times \quad 3 \\
\hline
\end{array}
$$

3
$$
\begin{array}{r}
2874 \\
\times \quad 5 \\
\hline
\end{array}
$$

4
$$
\begin{array}{r}
5742 \\
\times \quad 6 \\
\hline
\end{array}
$$

5
$$
\begin{array}{r}
7469 \\
\times \quad 4 \\
\hline
\end{array}
$$

6
$$
\begin{array}{r}
6310 \\
\times \quad 8 \\
\hline
\end{array}
$$

7 2,367 people visit a museum in the first week it opens. The same number come the second week. During the third week, 1,983 people visit. Circle the expression that shows how to find the total number of people who visited the museum in the first three weeks.

$(2 \times 2367) + 1983$ $\qquad$ $2 \times (2367 + 1983)$

$(2 \times 1983) + 2367$ $\qquad$ $2 \times (1983 + 2367)$

Multiplying a Two-Digit Number by a Two-Digit Number

You can multiply two-digit numbers using arrays.

Example

A nursery sells young trees. There are 12 rows of trees. There are 23 trees in each row. How many trees are there?

Step 1: Draw an array for 12 × 23. Separate each factor into tens and ones. Color each section a different color.

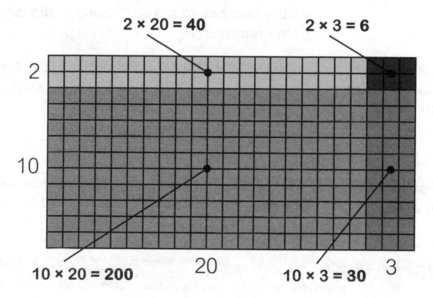

2 × 20 = 40   2 × 3 = 6

10 × 20 = 200   20   10 × 3 = 30   3

Step 2: The number of squares in each section shows partial products. Add the partial products. 200 + 30 + 40 + 6 = 276

The nursery has 276 trees.

Multiply

Find each product. On a separate sheet of paper, draw arrays to show partial products.

1 15 × 19 = _____285_____

2 22 × 17 = _____

3 53 × 26 = _____

4 11 × 14 = _____

5 25 × 12 = _____

6 8 × 18 = _____

Using Mental Math to Multiply

You can multiply a number by multiples of 10 or 100 using mental math.

Examples

A large market sells an average of 200 apples, 500 cherries, and 50 loaves of bread each day. How many apples does the market sell in 20 days? How many cherries does the market sell in 100 days? How many loaves of bread does the market sell in 400 days?

You can use a pattern to multiply 20 × 200.	The number of zeros in the product is the same as the number of zeros in the factors.	If the product of a fact ends in zero, then add a zero to the digits in the final product.
$20 × 2 = 40$ $20 × 20 = 400$ $20 × 200 = 4,000$	$500 × 100$ 4 zeros $50,000$ 4 zeros	$5 × 4 = 20$ $50 × 400 = 20,000$
The market sells 4,000 apples in 20 days.	The market sells 50,000 cherries in 100 days.	The market sells 20,000 loaves of bread in 400 days.

Multiply

Find each product using mental math.

1 $10 × 300 =$ ___3000___   **2** $40 × 90 =$ _____   **3** $400 × 30 =$ _____

4 $800 × 70 =$ _____   **5** $500 × 40 =$ _____   **6** $25 × 600 =$ _____

7 40 has one zero. 500 has two zeros. Why does the product of 40 and 500 have four zeros? Explain.

_____

_____

_____

Estimating Products

You already know how to round numbers to estimate. You can use rounding to estimate products too. You can also use compatible numbers to estimate.

Examples

A new hippopotamus named Pepe arrives at The San Diego Zoo. He eats 616 pounds of food every week. There are 52 weeks in a year. About how much food will Pepe eat in a year?

You can use rounding to estimate the amount of food Pepe eats in a year.

52 × 616
Round 616 to the nearest hundred.
52 × 600 = 31,200

Pepe eats about 31,200 pounds of food in a year.

You can use compatible numbers to estimate the amount of food Pepe eats in a year. Compatible numbers are easy to multiply.

52 × 616
Change 52 to 50.
Change 616 to 600.
50 × 600 = 30,000

Pepe eats about 30,000 pounds of food in a year.

Multiply

Use rounding or compatible numbers to estimate each product.

1 41 × 63 = _____2,400_____

4 81 × 33 = _____

2 53 × 51 = _____

5 36 × 22 = _____

3 19 × 84 = _____

6 32 × 37 = _____

7 An adult giraffe at the zoo eats 517 pounds of leaves in a week. Write the expression that shows the best way to estimate how many pounds of leaves the giraffe will eat in 5 weeks.

_____

Name _____

Problem Solving: Reasonable Answers

After you solve a problem, you can check your answer to see if it is reasonable.

Example

Connor collects rocks. He arranged his rocks in 6 rows with 23 rocks in each row. How many rocks are in Connor's rock collection?

When you solve the problem, ask yourself: Is the calculation reasonable?

There are 138 rocks in all.

Estimate: 6 × 20 = 120

The answer is reasonable because 120 is close to 138.

If there were 196 rocks in all.

Estimate: 6 × 20 = 120

The answer is not reasonable because 120 is not close to 196.

Solve

Make an estimate to show that each answer is reasonable.

1 Ms. Hunter spends $42 a week on groceries. How much money will she spend in 52 weeks?

Estimate: $2,000; Answer: $2,184

So far this year, Ms. Hunter has spent $1,324 on groceries. How much more will she spend this year?

_____

2 There are 12 large cages in a pet store. The owner puts 18 birds in each cage. How many birds are in the pet store?

_____

3 A children's library room has 6,729 children's nonfiction books, 4,902 picture books, and 5,886 children's fiction books. How many children's books does the library have in all?

_____

Name _____

Find each product using partial products. You can draw pictures for each problem on another sheet of paper.

1 16
× 4

2 58
× 6

3 97
× 8

4 18
× 3

5 63
× 5

6 55
× 9

Find each product.

7 266
× 6

8 793
× 5

9 625
× 8

10 4905
× 4

11 7990
× 8

12 5914
× 3

13 532
× 6

14 1715
× 8

15 9023
× 4

Find each product. You can draw arrays for each problem on another sheet of paper.

16 42 × 25 = _____

17 13 × 29 = _____

18 71 × 33 = _____

19 17 × 31 = _____

20 24 × 53 = _____

21 49 × 14 = _____

Name _____

Find each product using mental math.

22 400 × 50 = _____   **23** 30 × 600 = _____   **24** 70 × 400 = _____

Use rounding or compatible numbers to estimate each product.

25 52 × 61 = _____   **26** 47 × 11 = _____   **27** 897 × 9 = _____

Solve. Make an estimate to show that each answer is reasonable.

28 A hotel has 239 beds. There are 4 blankets for each bed. How many blankets does the hotel have?

_____

The hotel staff washes the blankets. So far, they have washed 402 blankets. How many blankets are left to wash?

_____

29 A used car lot has 16 rows of cars to sell. There are 12 cars in each row. How many cars are for sale in the used car lot?

_____

30 Marco has 31 green marbles, 19 red marbles, and 22 blue marbles. How many marbles does he have in all?

_____

Nicole has 209 marbles. How many more marbles does Nicole have than Marco?

_____

31 One city has a population of 174,533. A nearby city has 129,054 people. How many people live in the two communities.

_____

Dividing a Two-Digit Number

You can divide a two-digit number by a one-digit number.

Example

Mr. Jacobs has 84 student paintings. He wants to show the paintings on the walls of his classroom. If he divides the paintings equally, how many paintings should he put on each wall?

Draw a picture to show 84. Divide the tens into four equal groups.

Divide the ones.

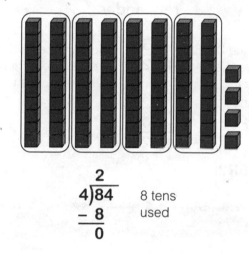

$$\begin{array}{r} 2 \\ 4\overline{)84} \\ -8 \\ \hline 0 \end{array}$$ 8 tens used

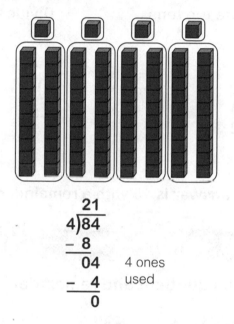

$$\begin{array}{r} 21 \\ 4\overline{)84} \\ -8 \\ \hline 04 \\ -4 \\ \hline 0 \end{array}$$ 4 ones used

Mr. Jacobs should put 21 paintings on each wall.

Divide

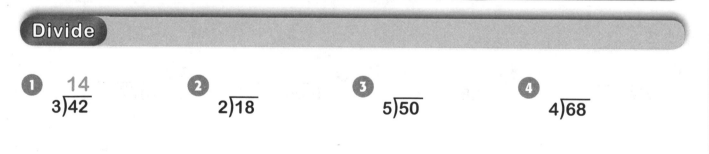

1 $\begin{array}{r} 14 \\ 3\overline{)42} \end{array}$

2 $2\overline{)18}$

3 $5\overline{)50}$

4 $4\overline{)68}$

5 Anwar has 42 books. He places 14 books in each box. Draw a picture to determine how many boxes Anwar used.

_____

Dividing a Two-Digit Number with Remainders

Sometimes you will have some amount left over after you divide. The amount left over is called the remainder.

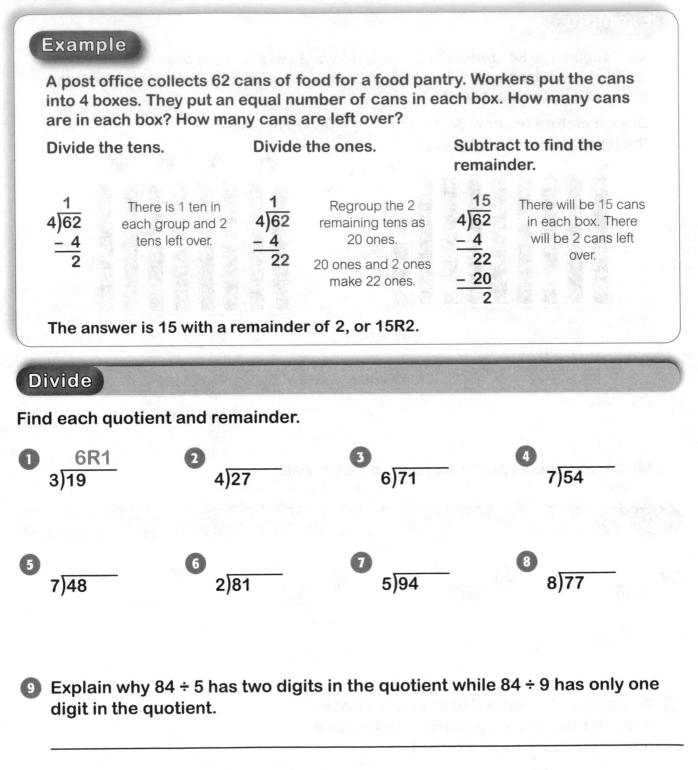

Example

A post office collects 62 cans of food for a food pantry. Workers put the cans into 4 boxes. They put an equal number of cans in each box. How many cans are in each box? How many cans are left over?

Divide the tens.

$$\begin{array}{r} 1 \\ 4\overline{)62} \\ -\underline{4} \\ 2 \end{array}$$

There is 1 ten in each group and 2 tens left over.

Divide the ones.

$$\begin{array}{r} 1 \\ 4\overline{)62} \\ -\underline{4} \\ 22 \end{array}$$

Regroup the 2 remaining tens as 20 ones.

20 ones and 2 ones make 22 ones.

Subtract to find the remainder.

$$\begin{array}{r} 15 \\ 4\overline{)62} \\ -\underline{4} \\ 22 \\ -\underline{20} \\ 2 \end{array}$$

There will be 15 cans in each box. There will be 2 cans left over.

The answer is 15 with a remainder of 2, or 15R2.

Divide

Find each quotient and remainder.

1
$$\begin{array}{r} 6R1 \\ 3\overline{)19} \end{array}$$

2
$$4\overline{)27}$$

3
$$6\overline{)71}$$

4
$$7\overline{)54}$$

5
$$7\overline{)48}$$

6
$$2\overline{)81}$$

7
$$5\overline{)94}$$

8
$$8\overline{)77}$$

9 Explain why 84 ÷ 5 has two digits in the quotient while 84 ÷ 9 has only one digit in the quotient.

_____

_____

Dividing a Three-Digit Number

Examples

A factory produces 137 balls of yarn on Monday. They ship the yarn in 4 boxes. The balls of yarn are equally divided. How many balls of yarn are in each box? How many are left over?

Ask yourself: Are there enough hundreds to divide? If not, start by dividing the tens.

$$\begin{array}{r} 3 \\ 4\overline{)137} \\ -12 \\ \hline 1 \end{array}$$

There are not enough hundreds. Start with the tens.

Divide the ones.

$$\begin{array}{r} 34 \\ 4\overline{)137} \\ -12 \\ \hline 17 \\ -16 \\ \hline 1 \end{array}$$

34 balls of yarn are in each box, with 1 ball of yarn left over

On Tuesday, the factory produces 297 balls of yarn and ships them in only 2 boxes. How many balls of yarn are in each box?

Ask yourself: Are there enough hundreds to divide? If so, divide the hundreds.

$$\begin{array}{r} 1 \\ 2\overline{)297} \\ -2 \\ \hline \end{array}$$

Divide the tens and ones.

$$\begin{array}{r} 148 \\ 2\overline{)297} \\ -2 \\ \hline 9 \\ -8 \\ \hline 17 \\ -16 \\ \hline 1 \end{array}$$

148 balls of yarn are in each box, with 1 ball of yarn left over

Divide

Find each quotient and remainder. Not all problems will have remainders.

1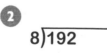

$$\begin{array}{r} 37R1 \\ 6\overline{)223} \end{array}$$

2

$$8\overline{)192}$$

3

$$4\overline{)162}$$

4 A large jar contains 321 buttons. Ms. Patel divides the buttons into 3 equal groups. Write the number sentence that shows how Ms. Patel divides the buttons.

_____

Name _____

Dividing a Four-Digit Number

Example

A bookstore has 3,491 books. The workers put the books in 3 large bookcases. There are an equal number of books in each bookcase. How many books are in each bookcase? How many books are left over?

Are there enough thousands? If so, divide the thousands first. If not, divide the hundreds.

```
    1
3)3491
  - 3
    1
```

Divide the hundreds, tens, and ones.

```
   1163
3)3491
 - 3
   4
 - 3
   19
 - 18
   11
  - 9
    2
```

You can multiply the quotient by the divisor and add the remainder to check your answer. Multiplication and division are inverse operations.

```
      1
   1163
 ×    3
   3489
```

3489 + 2 = 3491

The answer is 1163 with a remainder of 2, or 1163R2

Divide

Find each quotient. Use multiplication to check your answer.

1.
```
  1360R2
4)5442
```

2.
```
7)3988
```

3.
```
2)5842
```

4.
```
6)8173
```

5. A company bought 7 laptop computers for $3,234. Each computer was the same price. How much did 1 computer cost?

_____

Using Mental Math to Divide

You can use patterns to help you divide mentally.

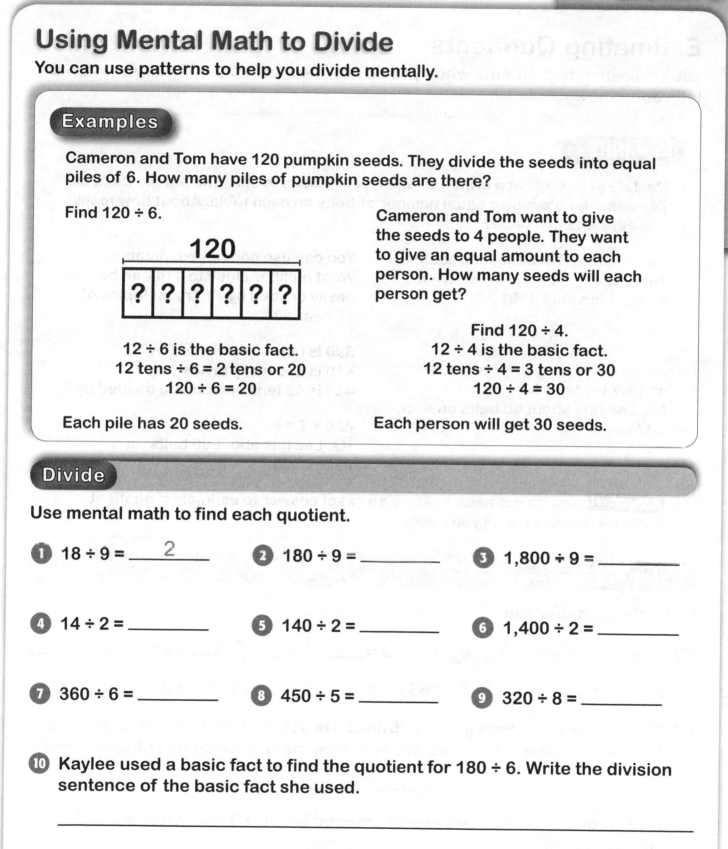

Examples

Cameron and Tom have 120 pumpkin seeds. They divide the seeds into equal piles of 6. How many piles of pumpkin seeds are there?

Find 120 ÷ 6.

120

| ? | ? | ? | ? | ? | ? |

12 ÷ 6 is the basic fact.
12 tens ÷ 6 = 2 tens or 20
120 ÷ 6 = 20

Each pile has 20 seeds.

Cameron and Tom want to give the seeds to 4 people. They want to give an equal amount to each person. How many seeds will each person get?

Find 120 ÷ 4.
12 ÷ 4 is the basic fact.
12 tens ÷ 4 = 3 tens or 30
120 ÷ 4 = 30

Each person will get 30 seeds.

Divide

Use mental math to find each quotient.

1 18 ÷ 9 = ___2___

2 180 ÷ 9 = _____

3 1,800 ÷ 9 = _____

4 14 ÷ 2 = _____

5 140 ÷ 2 = _____

6 1,400 ÷ 2 = _____

7 360 ÷ 6 = _____

8 450 ÷ 5 = _____

9 320 ÷ 8 = _____

10 Kaylee used a basic fact to find the quotient for 180 ÷ 6. Write the division sentence of the basic fact she used.

_____

Name _____

Estimating Quotients

You can estimate quotients when you do not need an exact answer. An estimate tells about how much.

Examples

Ms. Lee sells belts at a craft fair. She has 414 belts to sell. She places them on 7 tables. She places an equal number of belts on each table. About how many belts are on each table?

First, round 414 to the nearest ten. Then ask yourself: "7 times what number is about 410?"

$$7 \times 6 = 42$$
$$7 \times 60 = 420$$

$414 \div 7$ is about 60.
Ms. Lee has about 60 belts on each table.

You can use compatible numbers. What number close to 414 can be easily divided by 7? Try multiples of 10 near 400.

390 is not easily divided by 7.
410 is not easily divided by 7.
420 is 42 tens and can be divided by 7.

$$420 \div 7 = 60$$
Ms. Lee has about 60 belts on each table.

Remember! You do not need to know an exact answer to estimate a quotient. A rounded answer is all you need.

Estimate

Estimate each quotient.

1. $128 \div 3 =$ _about 40_

2. $132 \div 5 =$ _____

3. $652 \div 6 =$ _____

4. $252 \div 9 =$ _____

5. $765 \div 3 =$ _____

6. $742 \div 8 =$ _____

7. Mr. Freeman sold hats for a fundraiser. He sold 52 hats in 4 weeks. He sold the same number of hats each week. How many hats did he sell each week?

_____

Is an estimate or an exact answer needed to solve the problem above? How do you know?

_____

Name _____

Problem Solving: Extra or Missing Information

Some problems have information you do not need to solve the problem. Other problems are missing information and cannot be solved.

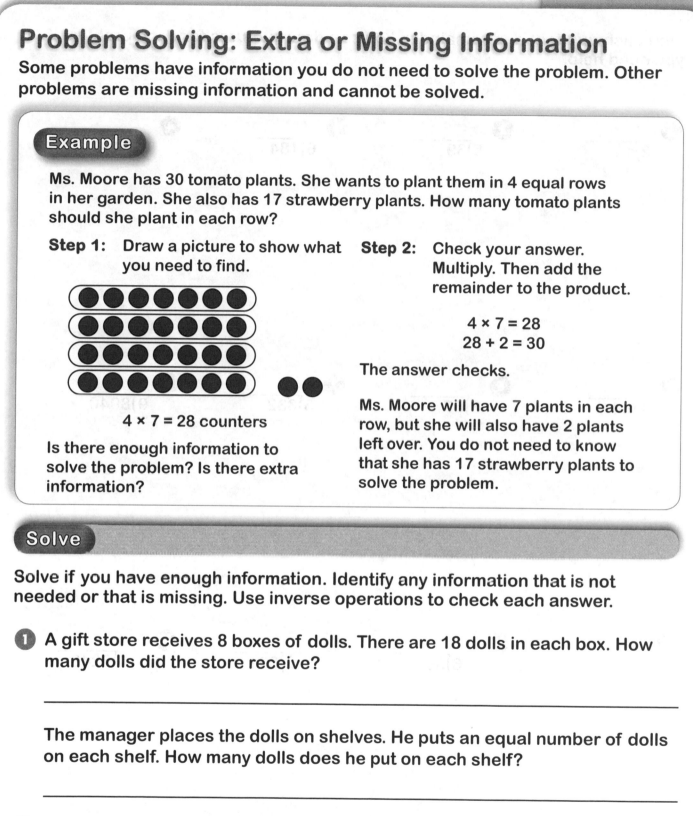

Example

Ms. Moore has 30 tomato plants. She wants to plant them in 4 equal rows in her garden. She also has 17 strawberry plants. How many tomato plants should she plant in each row?

Step 1: Draw a picture to show what you need to find.

$4 \times 7 = 28$ counters

Is there enough information to solve the problem? Is there extra information?

Step 2: Check your answer. Multiply. Then add the remainder to the product.

$$4 \times 7 = 28$$
$$28 + 2 = 30$$

The answer checks.

Ms. Moore will have 7 plants in each row, but she will also have 2 plants left over. You do not need to know that she has 17 strawberry plants to solve the problem.

Solve

Solve if you have enough information. Identify any information that is not needed or that is missing. Use inverse operations to check each answer.

1 A gift store receives 8 boxes of dolls. There are 18 dolls in each box. How many dolls did the store receive?

_____

The manager places the dolls on shelves. He puts an equal number of dolls on each shelf. How many dolls does he put on each shelf?

_____

2 A family has some money for a trip. If they travel for 4 days, how much money can they spend each day?

_____

Find each quotient and remainder. You can draw pictures for each problem if you need help.

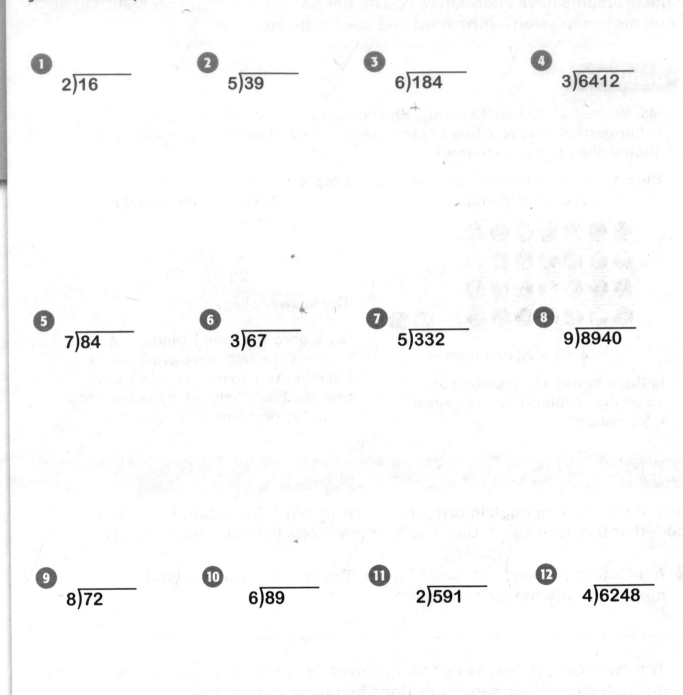

1. 2)16‾‾‾

2. 5)39‾‾‾

3. 6)184‾‾‾

4. 3)6412‾‾‾

5. 7)84‾‾‾

6. 3)67‾‾‾

7. 5)332‾‾‾

8. 9)8940‾‾‾

9. 8)72‾‾‾

10. 6)89‾‾‾

11. 2)591‾‾‾

12. 4)6248‾‾‾

Use mental math to find each quotient.

13 600 ÷ 6 = _____ **14** 540 ÷ 9 = _____ **15** 320 ÷ 4 = _____

Estimate each quotient.

16 181 ÷ 3 = _____ **17** 452 ÷ 5 = _____ **18** 876 ÷ 8 = _____

Solve. If you divide, explain what the remainder shows.

19 A deli has 5 trays of sandwiches. There are 6 sandwiches on each tray. How many sandwiches does the deli have?

_____

The manager places the sandwiches on 4 shelves. He puts an equal number of sandwiches on each shelf. How many sandwiches does he put on each shelf?

_____

20 Scott has 348 comic books. He places the comic books into 8 boxes. He puts an equal number of comic books in each box. How many comic books are in each box?

_____

21 639 people see a movie on Saturday. 711 people see the movie on Sunday. How many people saw the movie in all?

_____

427 people bought popcorn during the movie. How many people did not buy popcorn during the movie?

_____

22 Ms. Singh makes snacks for a party. She has some crackers and 6 paper cups. She puts 25 crackers into each cup. There are no crackers left over. Write an equation that shows this problem. Use the variable x for the total number of crackers.

_____

Name _____

Fractions

Fractions are symbols, such as $\frac{1}{2}$ or $\frac{2}{3}$. You can use fractions to name a part of a whole or parts of a set.

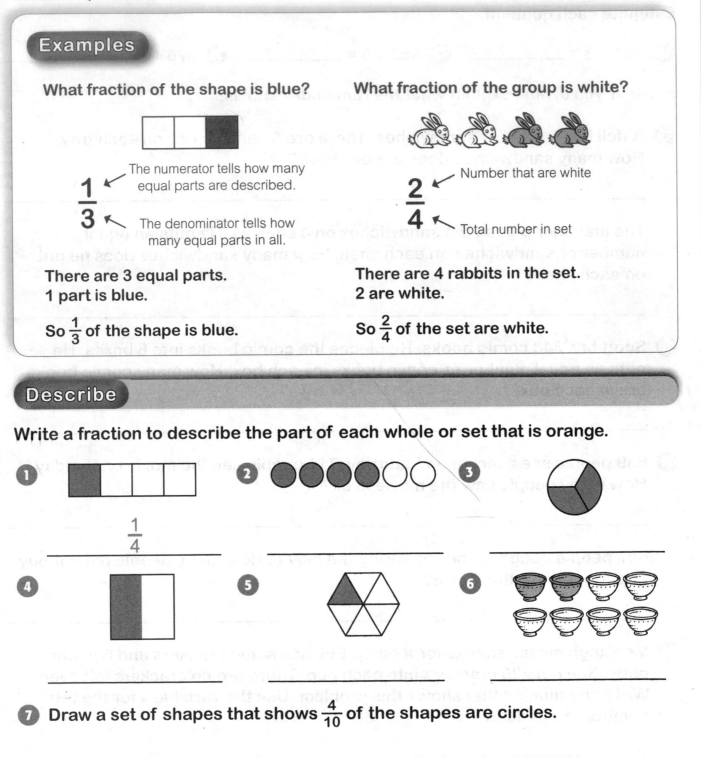

Examples

What fraction of the shape is blue?

$\frac{1}{3}$

The numerator tells how many equal parts are described.

The denominator tells how many equal parts in all.

There are 3 equal parts.
1 part is blue.

So $\frac{1}{3}$ of the shape is blue.

What fraction of the group is white?

$\frac{2}{4}$

Number that are white

Total number in set

There are 4 rabbits in the set.
2 are white.

So $\frac{2}{4}$ of the set are white.

Describe

Write a fraction to describe the part of each whole or set that is orange.

1. $\frac{1}{4}$

2. _____

3. _____

4. _____

5. _____

6. _____

7. Draw a set of shapes that shows $\frac{4}{10}$ of the shapes are circles.

Equivalent Fractions

Some fractions name the same part of a whole. They are equivalent fractions.

Example

Both $\frac{1}{3}$ and $\frac{2}{6}$ show the same part of the whole.

$\frac{1}{3}$ and $\frac{2}{6}$ are equivalent fractions.

1					
$\frac{1}{3}$		$\frac{1}{3}$		$\frac{1}{3}$	
$\frac{1}{6}$	$\frac{1}{6}$	$\frac{1}{6}$	$\frac{1}{6}$	$\frac{1}{6}$	$\frac{1}{6}$

Identify

Use the fraction strips to identify equivalent fractions.

1 What fraction is equivalent to $\frac{1}{3}$?

$\frac{2}{6}$

_____

2 What fraction is equivalent to $\frac{2}{8}$?

_____

3 What fraction is equivalent to $\frac{4}{6}$?

_____

4 What fraction is equivalent to $\frac{4}{5}$?

_____

1									
$\frac{1}{2}$					$\frac{1}{2}$				
$\frac{1}{3}$			$\frac{1}{3}$			$\frac{1}{3}$			
$\frac{1}{4}$		$\frac{1}{4}$		$\frac{1}{4}$			$\frac{1}{4}$		
$\frac{1}{5}$		$\frac{1}{5}$	$\frac{1}{5}$		$\frac{1}{5}$		$\frac{1}{5}$		
$\frac{1}{6}$	$\frac{1}{6}$	$\frac{1}{6}$		$\frac{1}{6}$		$\frac{1}{6}$		$\frac{1}{6}$	
$\frac{1}{8}$	$\frac{1}{8}$	$\frac{1}{8}$	$\frac{1}{8}$	$\frac{1}{8}$	$\frac{1}{8}$	$\frac{1}{8}$	$\frac{1}{8}$		
$\frac{1}{10}$	$\frac{1}{10}$	$\frac{1}{10}$	$\frac{1}{10}$	$\frac{1}{10}$	$\frac{1}{10}$	$\frac{1}{10}$	$\frac{1}{10}$	$\frac{1}{10}$	$\frac{1}{10}$

5 Write two equivalent fractions for the area that is red.

_____

Name _____

Finding Equivalent Fractions

You can multiply or divide to find equivalent fractions.

Example

Both $\frac{1}{2}$ and $\frac{4}{8}$ show the same part of the whole. They are equivalent fractions.

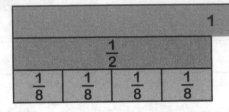

You can multiply to find an equivalent fraction.

$$\frac{1}{2}\frac{(\times 4)}{(\times 4)} = \frac{4}{8}$$

You can also divide.

$$\frac{4}{8}\frac{(\div 4)}{(\div 4)} = \frac{1}{2}$$

Multiply or Divide

Multiply or divide to find equivalent fractions.

1 $\frac{2}{3}\frac{(\times 5)}{(\times 5)} = \frac{10}{15}$

2 $\frac{6}{12}\frac{(\div 3)}{(\div 3)} = \frac{}{4}$

3 $\frac{8}{12} = \frac{4}{}$

4 $\frac{3}{4} = \frac{}{24}$

Find an equivalent fraction for each.

5 $\frac{3}{4} = $ _____

6 $\frac{1}{3} = $ _____

7 There are 30 students in a classroom. $\frac{1}{3}$ of their names begin with the letters M or C. How can you use equivalent fractions to find how many students this is?

_____

_____

Name _____

Comparing Fractions

You can compare fractions by finding equivalent fractions.

Example

Rita went to a carnival. She used $\frac{1}{2}$ of her money on rides and $\frac{2}{6}$ of her money on food. Did she spend more money on rides or food?

Step 1: Use an equivalent fraction.
What fraction is equivalent to $\frac{1}{2}$ and has a denominator of 6?

$$\frac{1}{2} \frac{(\times 3)}{(\times 3)} = \frac{3}{6}$$

Step 2: Compare the numerators.

$$\frac{3}{6} > \frac{2}{6}$$

$\frac{3}{6}$ is greater than $\frac{2}{6}$, so $\frac{1}{2}$ is greater than $\frac{2}{6}$.

Rita spent more money on rides.

Compare

Compare each pair of fractions. Write <, >, or =.

1. $\frac{2}{5}$ $>$ $\frac{1}{10}$

2. $\frac{1}{3}$ ___ $\frac{3}{6}$

3. $\frac{2}{4}$ ___ $\frac{1}{2}$

4. $\frac{6}{12}$ ___ $\frac{3}{3}$

5. $\frac{4}{6}$ ___ $\frac{2}{3}$

6. $\frac{2}{3}$ ___ $\frac{5}{6}$

7. $\frac{4}{9}$ ___ $\frac{1}{3}$

8. $\frac{3}{5}$ ___ $\frac{4}{15}$

9. $\frac{1}{2}$ ___ $\frac{7}{8}$

10. Alexis drew the picture on the right to show that $\frac{3}{4}$ is less than $\frac{4}{8}$. What was her mistake?

_____

_____

Name _____

Using Benchmarks to Compare Fractions

You can use benchmark fractions to compare fractions too. Drawings can help.

Example

Josh and Petra are reading the same book. So far, Josh has read $\frac{1}{3}$ of the book. Petra has read $\frac{3}{5}$. Who has read more?

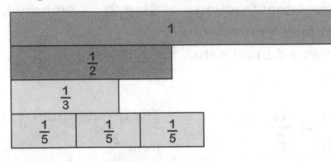

You can use drawings to compare $\frac{1}{3}$ and $\frac{3}{5}$ to a benchmark fraction.

A good benchmark is $\frac{1}{2}$.

$$\frac{1}{3} < \frac{1}{2} \qquad \frac{3}{5} > \frac{1}{2}$$

$$\text{So } \frac{3}{5} > \frac{1}{3}.$$

Petra has read more of the book.

Compare

Compare each pair of fractions. Write <, >, or =. Use $\frac{1}{2}$ or $\frac{1}{4}$ as benchmarks.

1. $\frac{1}{3}$ __<__ $\frac{3}{4}$

2. $\frac{2}{5}$ ___ $\frac{3}{8}$

3. $\frac{2}{4}$ ___ $\frac{2}{6}$

4. $\frac{2}{3}$ ___ $\frac{3}{6}$

5. $\frac{1}{3}$ ___ $\frac{2}{6}$

6. $\frac{4}{8}$ ___ $\frac{5}{8}$

7. A school flag is $\frac{3}{8}$ gold and $\frac{1}{5}$ brown. Is more of the flag colored gold or brown?

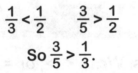

Identifying Valid Comparisons

When you compare fractions, you must be sure they refer to the same whole. Sometimes drawings can help.

Example

Jed draws the squares below. They both show $\frac{1}{2}$. He says that since both squares show $\frac{1}{2}$, they show the same amount. Is he correct?

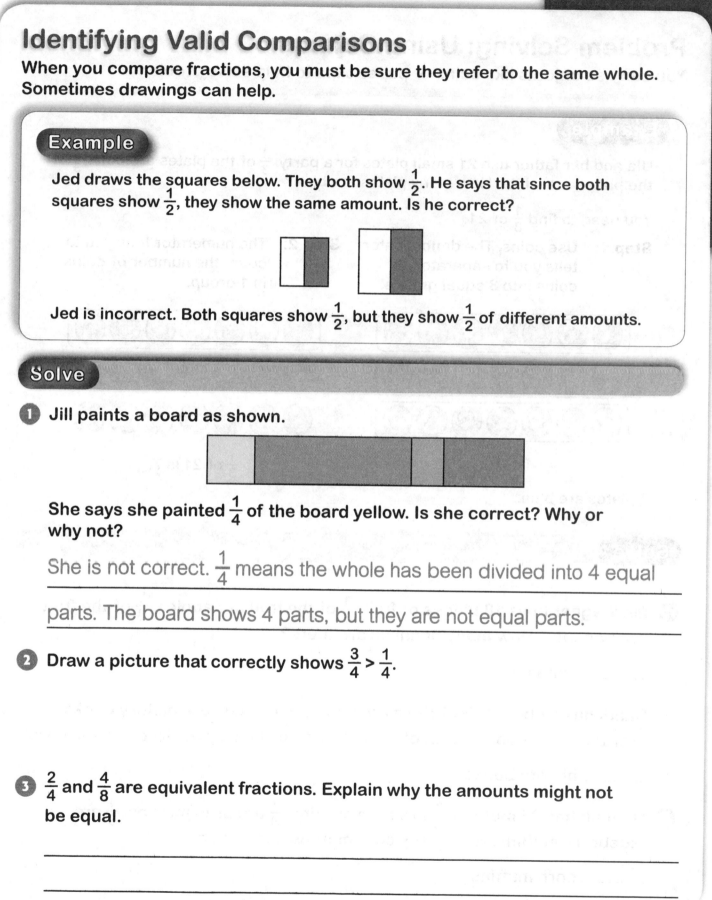

Jed is incorrect. Both squares show $\frac{1}{2}$, but they show $\frac{1}{2}$ of different amounts.

Solve

1 Jill paints a board as shown.

She says she painted $\frac{1}{4}$ of the board yellow. Is she correct? Why or why not?

She is not correct. $\frac{1}{4}$ means the whole has been divided into 4 equal

parts. The board shows 4 parts, but they are not equal parts.

2 Draw a picture that correctly shows $\frac{3}{4} > \frac{1}{4}$.

3 $\frac{2}{4}$ and $\frac{4}{8}$ are equivalent fractions. Explain why the amounts might not be equal.

_____

_____

Lesson 7

Problem Solving: Using Objects

You can use objects like coins or paper clips to help solve problems.

Example

Ula and her father use 21 small plates for a party. $\frac{1}{3}$ of the plates are blue. $\frac{2}{3}$ of the plates are white. How many plates are blue?

You need to find $\frac{1}{3}$ of 21.

Step 1: Use coins. The denominator tells you to separate the coins into 3 equal groups.

Step 2: The numerator tells you to count the number of coins in 1 group.

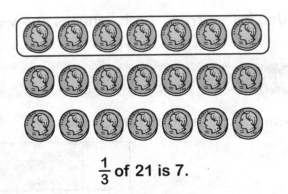

$\frac{1}{3}$ of 21 is 7.

7 plates are blue.

Solve

1. Mr. Wagner buys 18 pieces of fruit. $\frac{1}{3}$ of the fruit are apples. $\frac{2}{3}$ of the fruit are bananas. How many bananas are there?

   ___12___ bananas

2. Sandi has 12 books. $\frac{2}{4}$ of them are fiction. $\frac{1}{4}$ of them are history books. $\frac{1}{4}$ of them are books about planes. How many history books does she have?

   _____ history books

3. Hannah has 24 muffins. $\frac{1}{3}$ are corn muffins. $\frac{1}{3}$ are bran muffins. $\frac{1}{3}$ are blueberry muffins. How many corn muffins does she have?

   _____ corn muffins

Name _____

Write a fraction to describe the part of each whole or set that is green.

1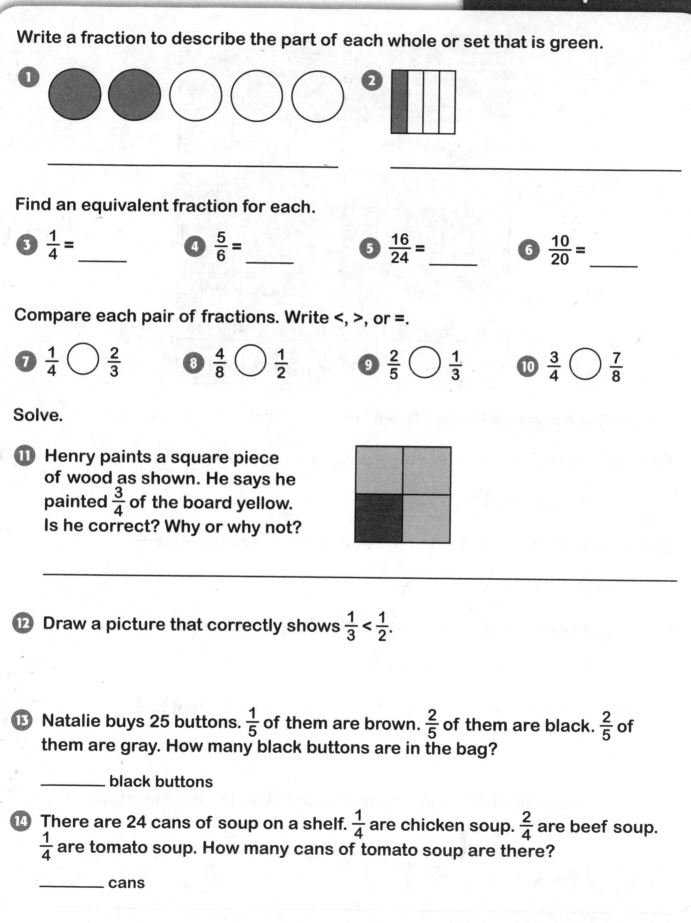

2

_____   _____

Find an equivalent fraction for each.

3 $\frac{1}{4}$ = _____

4 $\frac{5}{6}$ = _____

5 $\frac{16}{24}$ = _____

6 $\frac{10}{20}$ = _____

Compare each pair of fractions. Write <, >, or =.

7 $\frac{1}{4}$ ◯ $\frac{2}{3}$

8 $\frac{4}{8}$ ◯ $\frac{1}{2}$

9 $\frac{2}{5}$ ◯ $\frac{1}{3}$

10 $\frac{3}{4}$ ◯ $\frac{7}{8}$

Solve.

11 Henry paints a square piece of wood as shown. He says he painted $\frac{3}{4}$ of the board yellow. Is he correct? Why or why not?

_____

12 Draw a picture that correctly shows $\frac{1}{3} < \frac{1}{2}$.

13 Natalie buys 25 buttons. $\frac{1}{5}$ of them are brown. $\frac{2}{5}$ of them are black. $\frac{2}{5}$ of them are gray. How many black buttons are in the bag?

_____ black buttons

14 There are 24 cans of soup on a shelf. $\frac{1}{4}$ are chicken soup. $\frac{2}{4}$ are beef soup. $\frac{1}{4}$ are tomato soup. How many cans of tomato soup are there?

_____ cans

Name _____

		1		

$\frac{1}{2}$	$\frac{1}{2}$

$\frac{1}{3}$	$\frac{1}{3}$	$\frac{1}{3}$

$\frac{1}{4}$	$\frac{1}{4}$	$\frac{1}{4}$	$\frac{1}{4}$

$\frac{1}{5}$	$\frac{1}{5}$	$\frac{1}{5}$	$\frac{1}{5}$	$\frac{1}{5}$

$\frac{1}{6}$	$\frac{1}{6}$	$\frac{1}{6}$	$\frac{1}{6}$	$\frac{1}{6}$	$\frac{1}{6}$

$\frac{1}{8}$	$\frac{1}{8}$	$\frac{1}{8}$	$\frac{1}{8}$	$\frac{1}{8}$	$\frac{1}{8}$	$\frac{1}{8}$	$\frac{1}{8}$

$\frac{1}{10}$	$\frac{1}{10}$	$\frac{1}{10}$	$\frac{1}{10}$	$\frac{1}{10}$	$\frac{1}{10}$	$\frac{1}{10}$	$\frac{1}{10}$	$\frac{1}{10}$	$\frac{1}{10}$

Use the fraction strips to identify equivalent fractions.

15 Place a check mark next to the fraction that is equivalent to $\frac{2}{10}$.

_____ $\frac{1}{5}$ 　　_____ $\frac{3}{10}$ 　　_____ $\frac{1}{4}$

16 Place a check mark next to the fraction that is equivalent to $\frac{1}{3}$.

_____ $\frac{4}{8}$ 　　_____ $\frac{1}{2}$ 　　_____ $\frac{2}{6}$

17 Place a check mark next to the fraction that is equivalent to $\frac{6}{8}$.

_____ $\frac{9}{10}$ 　　_____ $\frac{3}{4}$ 　　_____ $\frac{4}{5}$

18 Place a check mark next to the fraction that is equivalent to $\frac{1}{2}$.

_____ $\frac{2}{3}$ 　　_____ $\frac{3}{8}$ 　　_____ $\frac{3}{6}$

Compare each pair of fractions. Write <, >, or =. Use the fraction strips and $\frac{1}{2}$ or $\frac{1}{4}$ as benchmarks.

19 $\frac{2}{3}$ ◯ $\frac{4}{8}$ 　　　　**20** $\frac{4}{5}$ ◯ $\frac{3}{4}$ 　　　　**21** $\frac{1}{5}$ ◯ $\frac{2}{10}$

Adding Fractions

You can add fractions. Sometimes you need to write the sums in simplest form.

Example

Frank likes to swim. He swims $\frac{1}{8}$ of a mile on Monday. He swims $\frac{3}{8}$ of a mile on Tuesday. How far does he swim in all?

Step 1: You can add numerators if denominators are the same.

$$\frac{1}{8} + \frac{3}{8} = \frac{1+3}{8} = \frac{4}{8}$$

You can also use drawings to help add fractions.

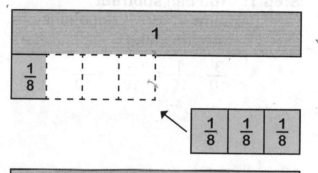

Step 2: Write your answer in simplest form. A fraction is in simplest form when 1 is the only number that divides both the denominator and the numerator with no remainder.

$$\frac{4}{8} \begin{smallmatrix}(\div 2)\\(\div 2)\end{smallmatrix} = \frac{2}{4} \begin{smallmatrix}(\div 2)\\(\div 2)\end{smallmatrix} = \frac{1}{2}$$

Frank swims $\frac{1}{2}$ of a mile.

Add

Write your answers in simplest form.

1 $\frac{2}{5} + \frac{2}{5} = \underline{\quad \frac{4}{5} \quad}$

2 $\frac{1}{3} + \frac{1}{3} = \underline{\qquad}$

3 $\frac{2}{8} + \frac{4}{8} = \underline{\qquad}$

4 Jeremy and Rosa are solving a math problem. They add $\frac{1}{3} + \frac{2}{3}$. Jeremy says the answer is $\frac{3}{3}$. Rosa says the answer is 1. Who is correct? Why?

_____

Name _____

Subtracting Fractions

You can subtract fractions too. Sometimes you need to write the differences in their simplest form.

Examples

Darrel buys $\frac{1}{10}$ pound of carrots. Karen buys $\frac{3}{10}$ pound of carrots. How much more does Karen buy than Darrel?

Step 1: You can subtract numerators if denominators are the same.

$$\frac{3}{10} - \frac{1}{10} = \frac{3-1}{10} = \frac{2}{10}$$

You can also use drawings to help subtract fractions.

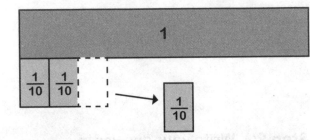

Step 2: Write your answer in simplest form.

$$\frac{2}{10} \begin{smallmatrix}(\div 2)\\[6pt](\div 2)\end{smallmatrix} = \frac{1}{5}$$

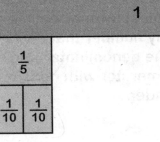

Karen buys $\frac{1}{5}$ pound more than Darrel.

Subtract

Write your answers in simplest form.

1. $\frac{3}{4} - \frac{2}{4} =$ _____ $\frac{1}{4}$ _____

2. $\frac{3}{5} - \frac{1}{5} =$ _____

3. $\frac{5}{6} - \frac{3}{6} =$ _____

4. $\frac{5}{8} - \frac{1}{8} =$ _____

5. $\frac{8}{10} - \frac{5}{10} =$ _____

6. $\frac{2}{3} - \frac{1}{3} =$ _____

7. Joel wants to solve $1 - \frac{3}{8}$. Explain how he can subtract.

_____

_____

Adding and Subtracting Fractions

Examples

Find $\frac{13}{15} - \frac{8}{15}$.

When you subtract fractions, you can subtract numerators if denominators are the same.

$$\frac{13}{15} - \frac{8}{15} = \frac{13-8}{15} = \frac{5}{15}$$

Write your answer in simplest form.

$$\frac{5 \,(\div 5)}{15 \,(\div 5)} = \frac{1}{3}$$

$$\frac{13}{15} - \frac{8}{15} = \frac{1}{3}$$

Find $\frac{5}{13} + \frac{6}{13}$.

When you add fractions, you can add numerators if denominators are the same.

$$\frac{5}{13} + \frac{6}{13} = \frac{5+6}{13} = \frac{11}{13}$$

Write your answer in simplest form.

$$\frac{11 \,(\div 1)}{13 \,(\div 1)} = \frac{11}{13}$$

$\frac{11}{13}$ is already in simplest form. 1 is the only number that divides both the numerator and the denominator with no remainder.

$$\frac{5}{13} + \frac{6}{13} = \frac{11}{13}$$

Add or Subtract

Write your answers in simplest form.

1. $\frac{4}{12} + \frac{4}{12} =$ ___ $\frac{2}{3}$ ___

2. $\frac{4}{5} - \frac{2}{5} =$ _____

3. $\frac{6}{10} + \frac{2}{10} =$ _____

4. $\frac{2}{4} + \frac{1}{4} =$ _____

5. $\frac{4}{7} - \frac{1}{7} =$ _____

6. $\frac{2}{8} + \frac{5}{8} =$ _____

7. $\frac{3}{5} + \frac{1}{5} =$ _____

8. $\frac{6}{9} + \frac{2}{9} =$ _____

9. $\frac{1}{11} + \frac{1}{11} =$ _____

10. Steve, Gabe, and Lena are decorating a wall. Steve decorates $\frac{3}{7}$ of the wall, Gabe decorates $\frac{1}{7}$, and Lena decorates $\frac{3}{7}$. How much more of the wall do they have to decorate?

_____

_____

Name _____

Mixed Numbers and Fractions

You can show all the equal parts in a fraction or a mixed number.

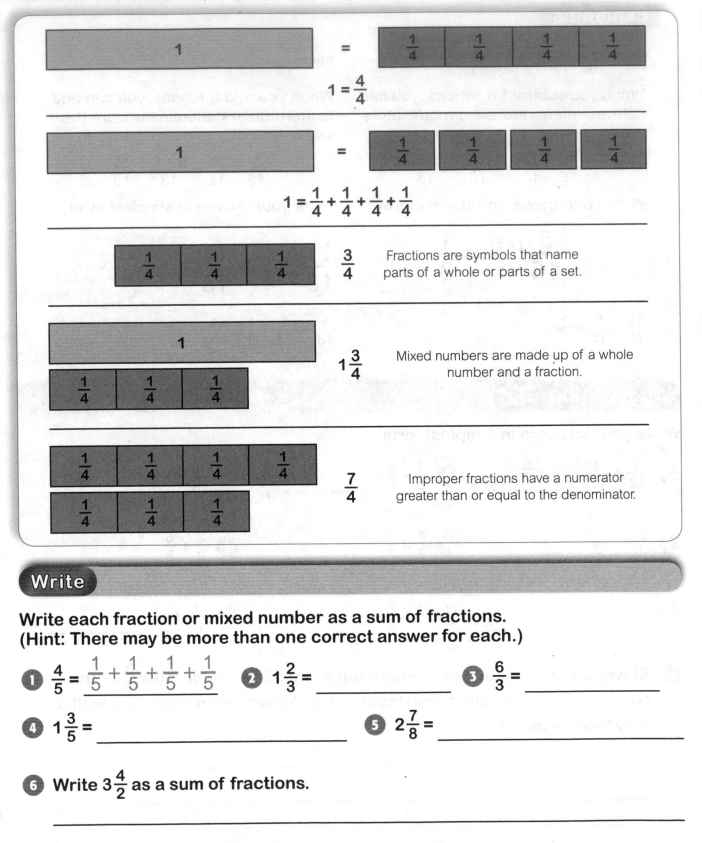

1	=	$\frac{1}{4}$ $\frac{1}{4}$ $\frac{1}{4}$ $\frac{1}{4}$

$$1 = \frac{4}{4}$$

1	=	$\frac{1}{4}$ $\frac{1}{4}$ $\frac{1}{4}$ $\frac{1}{4}$

$$1 = \frac{1}{4} + \frac{1}{4} + \frac{1}{4} + \frac{1}{4}$$

$\frac{1}{4}$ $\frac{1}{4}$ $\frac{1}{4}$ $\frac{3}{4}$ Fractions are symbols that name parts of a whole or parts of a set.

1 $\frac{1}{4}$ $\frac{1}{4}$ $\frac{1}{4}$ $1\frac{3}{4}$ Mixed numbers are made up of a whole number and a fraction.

$\frac{1}{4}$ $\frac{1}{4}$ $\frac{1}{4}$ $\frac{1}{4}$ $\frac{1}{4}$ $\frac{1}{4}$ $\frac{1}{4}$ $\frac{7}{4}$ Improper fractions have a numerator greater than or equal to the denominator.

Write

Write each fraction or mixed number as a sum of fractions.
(Hint: There may be more than one correct answer for each.)

1 $\frac{4}{5} = \frac{1}{5} + \frac{1}{5} + \frac{1}{5} + \frac{1}{5}$ **2** $1\frac{2}{3} =$ _____   **3** $\frac{6}{3} =$ _____

4 $1\frac{3}{5} =$ _____   **5** $2\frac{7}{8} =$ _____

6 Write $3\frac{4}{2}$ as a sum of fractions.

_____

Adding Mixed Numbers

You can add mixed numbers. Sometimes you need to write the sums in simplest form.

Example

Francine made a snack to eat while hiking. She used $2\frac{1}{4}$ cups of granola and $1\frac{3}{4}$ cups of raisins. How many cups of snack mix did she make?

Step 1: Add the fractions first.

$$
\begin{array}{r}
2\frac{1}{4} \\
+\ 1\frac{3}{4} \\
\hline
\frac{4}{4}
\end{array}
$$

Step 2: Add the whole numbers. Write your answer in simplest form.

$$
\begin{array}{r}
2\frac{1}{4} \\
+\ 1\frac{3}{4} \\
\hline
3\frac{4}{4} = 3 + 1 = 4
\end{array}
$$

Francine made 4 cups of snack mix.

Add

Write your answers in simplest form.

1
$$
\begin{array}{r}
1\frac{1}{2} \\
+\ 4\frac{1}{2} \\
\hline
6
\end{array}
$$

2
$$
\begin{array}{r}
4\frac{1}{3} \\
+\ 3\frac{1}{3} \\
\hline
\end{array}
$$

3
$$
\begin{array}{r}
16\frac{2}{7} \\
+\ 9\frac{4}{7} \\
\hline
\end{array}
$$

4
$$
\begin{array}{r}
24\frac{3}{5} \\
+\ 7\frac{1}{5} \\
\hline
\end{array}
$$

5 Last month Mr. Garrett flew three times. The first flight lasted $4\frac{4}{5}$ hours and the second flight lasted $8\frac{1}{5}$ hours. The third flight lasted $6\frac{3}{5}$ hours. For how many hours did Mr. Garrett travel?

_____

Name _____

Subtracting Mixed Numbers

You can subtract mixed numbers. Sometimes you need to write the differences in simplest form.

Example

Diane's bowl holds $3\frac{7}{8}$ cups of soup. She eats $2\frac{3}{8}$ cups at lunch and saves the rest for later. How much soup does she save for later?

Step 1: Subtract the fractions first.

$$3\frac{7}{8}$$
$$-\,2\frac{3}{8}$$
$$\overline{\frac{4}{8}}$$

Step 2: Subtract the whole numbers. Write your answer in simplest form.

$$3\frac{7}{8}$$
$$-\,2\frac{3}{8}$$
$$\overline{1\frac{4}{8}} = 1\frac{1}{2}$$

Diane saved $1\frac{1}{2}$ cups of soup for later.

Subtract

Write your answers in simplest form.

1 $5\frac{6}{8}$
$-\,2\frac{1}{8}$
$\overline{3\frac{5}{8}}$

2 $7\frac{5}{6}$
$-\,3\frac{2}{6}$

3 $8\frac{2}{5}$
$-\,2\frac{1}{5}$

4 $75\frac{3}{4}$
$-\,26\frac{3}{4}$

5 Olivia has 7 cups of water in a pitcher. She pours $2\frac{7}{8}$ cups of water into a glass. How can she use subtraction to find how much water is left in the pitcher?

_____

Adding and Subtracting Mixed Numbers

Examples

Find $4\frac{6}{9} - 1\frac{3}{9}$

When you subtract mixed numbers, you subtract the fractions first.

$$4\frac{6}{9}$$
$$-1\frac{3}{9}$$
$$\overline{\frac{3}{9}}$$

Then subtract the whole numbers. Write your answer in simplest form.

$$4\frac{6}{9}$$
$$-1\frac{3}{9}$$
$$\overline{3\frac{3}{9} = 1\frac{1}{3}}$$

Find $4\frac{2}{7} + 3\frac{4}{7}$

When you add mixed numbers, you add the fractions first.

$$4\frac{2}{7}$$
$$+3\frac{4}{7}$$
$$\overline{\frac{6}{7}}$$

Then add the whole numbers. Write your answer in simplest form.

$$4\frac{2}{7}$$
$$+3\frac{4}{7}$$
$$\overline{7\frac{6}{7} = 7\frac{6}{7}}$$

Add or Subtract

Add or subtract. Write your answers in simplest form.

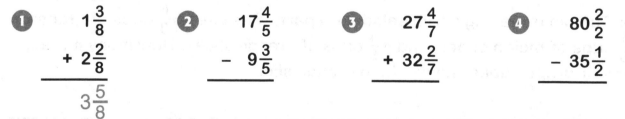

1 $1\frac{3}{8}$
     $+2\frac{2}{8}$
     $\overline{3\frac{5}{8}}$

2 $17\frac{4}{5}$
     $-9\frac{3}{5}$

3 $27\frac{4}{7}$
     $+32\frac{2}{7}$

4 $80\frac{2}{2}$
     $-35\frac{1}{2}$

5 The length of a baby Gila monster is $5\frac{7}{8}$ inches. The length of an adult is $21\frac{1}{8}$ inches. How much longer is the adult than the baby?

_____

Problem Solving: Draw a Picture and Write an Equation

Drawing pictures and writing equations are good ways to solve problems.

Example

A bus travels $23\frac{3}{5}$ miles in the morning and $31\frac{1}{5}$ miles in the afternoon. How many miles does the bus travel in all?

Step 1: Read the problem. What do you know?
The bus travels $23\frac{3}{5}$ miles and then $31\frac{1}{5}$ miles.

Step 2: Draw a picture.

? miles in all

$23\frac{3}{5}$	$31\frac{1}{5}$

Step 3: Write an equation. Add and simplify.

$$\begin{array}{r} 23\frac{3}{5} \\ + \ 31\frac{1}{5} \\ \hline 54\frac{4}{5} \end{array}$$

The bus travels $54\frac{4}{5}$ miles in all.

Solve

1. Ms. Robinson and her son hiked three trails with the following lengths: $5\frac{2}{9}$ mile; $1\frac{1}{9}$ mile; and $1\frac{4}{9}$ mile. How far did they walk in all?

$7\frac{7}{9}$ miles

2. Ms. Tanaka is making a fruit salad for a party. She uses $6\frac{6}{8}$ cups of grapes, $3\frac{1}{8}$ cups of melon cubes, and $4\frac{4}{8}$ cups of orange slices. How many more cups of grapes does she use than orange slices?

_____

3. A fabric store sold $8\frac{5}{9}$ yards of black felt and $4\frac{4}{9}$ yards of red felt on Tuesday. How many more yards of black felt than red felt were sold?

_____

Fractions, Repeated Addition, and Multiplication

Number sentences using repeated addition can be used to show multiplication of fractions.

Example

Each serving of punch has $\frac{1}{6}$ cup of orange juice. To make 9 servings of punch, how many cups of orange juice will you need?

Add $\frac{1}{6}$ a total of 9 times.

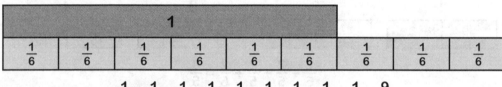

$$\frac{1}{6} + \frac{1}{6} + \frac{1}{6} + \frac{1}{6} + \frac{1}{6} + \frac{1}{6} + \frac{1}{6} + \frac{1}{6} + \frac{1}{6} = \frac{9}{6}$$

This is the same as multiplying $\frac{1}{6}$ by 9.

$$9 \times \frac{1}{6} = \frac{9}{6}$$

Then simplify your answer.

$$\frac{9}{6} = \frac{6}{6} + \frac{3}{6} = 1\frac{3}{6} = 1\frac{1}{2}$$

You need $1\frac{1}{2}$ cups of orange juice.

Interpret

Write an equation to show each fraction as a multiple.

1 $\frac{3}{4} = $ $\underline{\quad 3 \times \frac{1}{4} = \frac{3}{4} \quad}$

2 $\frac{4}{5} = $ _____

3 $\frac{3}{6} = $ _____

4 $\frac{2}{2} = $ _____

5 $\frac{5}{8} = $ _____

6 $\frac{4}{7} = $ _____

Name _____

Multiplying Fractions by Whole Numbers

You can multiply fractions by whole numbers using what you know about repeated addition.

Example

Jay is making trail mix for his friends. One serving contains $\frac{2}{5}$ cup of raisins. How many cups of raisins are in 4 servings?

Multiply $4 \times \frac{2}{5}$.

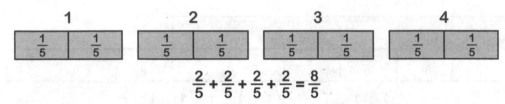

$$\frac{2}{5} + \frac{2}{5} + \frac{2}{5} + \frac{2}{5} = \frac{8}{5}$$

This is the same as multiplying $8 \times \frac{1}{5}$.

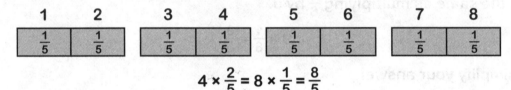

$$4 \times \frac{2}{5} = 8 \times \frac{1}{5} = \frac{8}{5}$$

Simplify your answer.

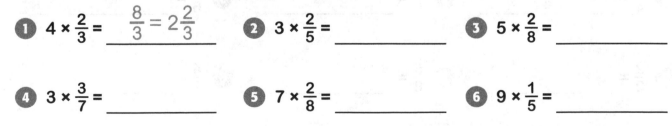

$$\frac{8}{5} = 1\frac{3}{5}$$

There are $1\frac{3}{5}$ cups of raisins in 4 servings.

Multiply

Write your answers in simplest form.

1 $4 \times \frac{2}{3} =$ _____ $\frac{8}{3} = 2\frac{2}{3}$

2 $3 \times \frac{2}{5} =$ _____

3 $5 \times \frac{2}{8} =$ _____

4 $3 \times \frac{3}{7} =$ _____

5 $7 \times \frac{2}{8} =$ _____

6 $9 \times \frac{1}{5} =$ _____

Problem Solving: Multiplying Whole Numbers by Fractions

Multiplying a whole number by a fraction can help you find a fraction of a whole number or a fraction of a set.

Example

Dan has a board that is 4 feet long. Nadeen has a board that is $\frac{2}{3}$ the length of Dan's. How long is Nadeen's board?

Step 1: Read the problem. What do you know? What are you asked to find?

Step 2: Find $4 \times \frac{2}{3}$. Draw a picture and write an equation.

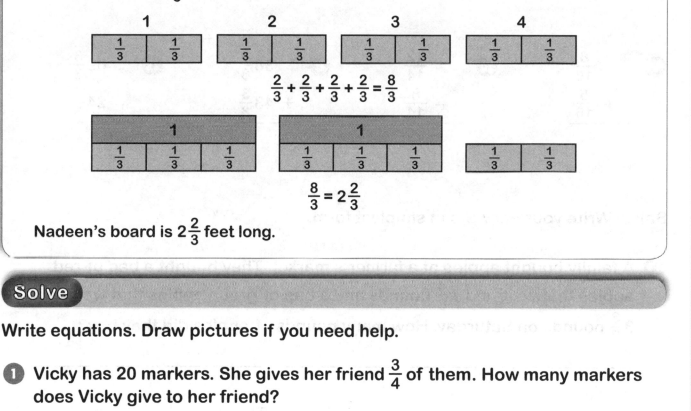

Nadeen's board is $2\frac{2}{3}$ feet long.

Solve

Write equations. Draw pictures if you need help.

1. Vicky has 20 markers. She gives her friend $\frac{3}{4}$ of them. How many markers does Vicky give to her friend?

_____

2. 36 students sign up for band class. $\frac{1}{3}$ of them want to play the saxophone. How many students want to play the saxophone?

_____

Name _____

Add or subtract. Write your answers in simplest form.

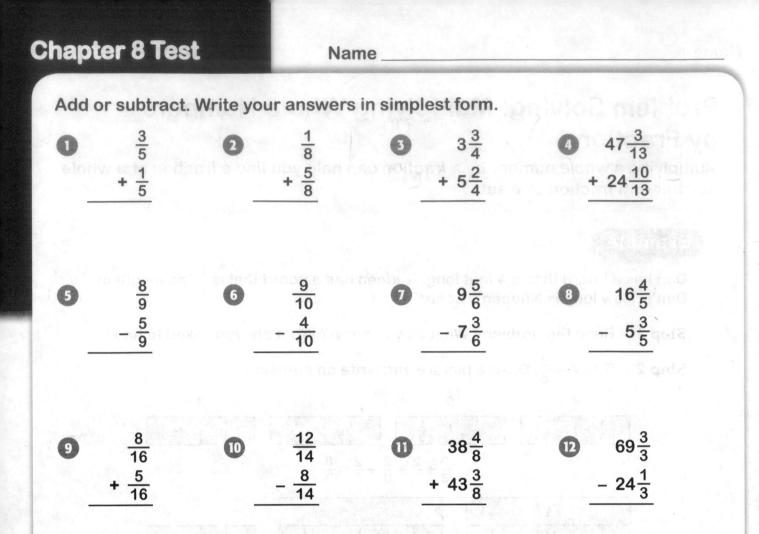

1. $\frac{3}{5}$
    $+ \frac{1}{5}$

2. $\frac{1}{8}$
    $+ \frac{5}{8}$

3. $3\frac{3}{4}$
    $+ 5\frac{2}{4}$

4. $47\frac{3}{13}$
    $+ 24\frac{10}{13}$

5. $\frac{8}{9}$
    $- \frac{5}{9}$

6. $\frac{9}{10}$
    $- \frac{4}{10}$

7. $9\frac{5}{6}$
    $- 7\frac{3}{6}$

8. $16\frac{4}{5}$
    $- 5\frac{3}{5}$

9. $\frac{8}{16}$
    $+ \frac{5}{16}$

10. $\frac{12}{14}$
    $- \frac{8}{14}$

11. $38\frac{4}{8}$
    $+ 43\frac{3}{8}$

12. $69\frac{3}{3}$
    $- 24\frac{1}{3}$

Solve. Write your answers in simplest form.

13. A family bought apples at a farmer's market. They bought a bag of red apples that weighed $7\frac{4}{8}$ pounds and a bag of green apples that weighed $3\frac{3}{8}$ pounds on Saturday. How many pounds of apples did they buy?

_____

14. Alex makes a salad for the family. He uses $4\frac{2}{9}$ cups of lettuce, $1\frac{1}{9}$ cups of tomatoes, and $2\frac{4}{9}$ cups of cucumbers. How many more cups of lettuce does he use than cucumbers?

_____

Solve. Write your answers in simplest form.

15 Write $\frac{4}{9}$ as a sum of fractions.

_____

16 Write $1\frac{3}{5}$ as a sum of fractions.

_____

Multiply. Write your answers in simplest form. Draw pictures if you need help.

17 $4 \times \frac{1}{3} =$ _____

19 $3 \times \frac{3}{5} =$ _____

18 $5 \times \frac{3}{4} =$ _____

20 $7 \times \frac{4}{8} =$ _____

Solve. Write your answers in simplest form.

21 Jason is practicing for a footrace. He runs $7\frac{3}{8}$ miles the first week and $9\frac{6}{8}$ miles the second week. How many miles does Jason run in all?

_____

22 76 vehicles are in a parking lot. $\frac{1}{4}$ of them are trucks. How many trucks are in the parking lot?

_____

23 A hardware store sold $7\frac{8}{10}$ feet of chain in one week. The store sold $4\frac{3}{10}$ feet of chain the next week. How many more feet of chain did the store sell in the first week?

_____

24 Linda travels to visit friends. She drives for a total of 23 miles. Damon also visits friends, but his trip is $\frac{2}{3}$ as long as Linda's. How many miles does Damon drive?

_____

Name _____

Denominators 10 and 100

When a fraction has a denominator of 10, you can multiply by 10 to find an equivalent fraction. When a fraction has a denominator of 100, you can divide by 10 to find an equivalent fraction.

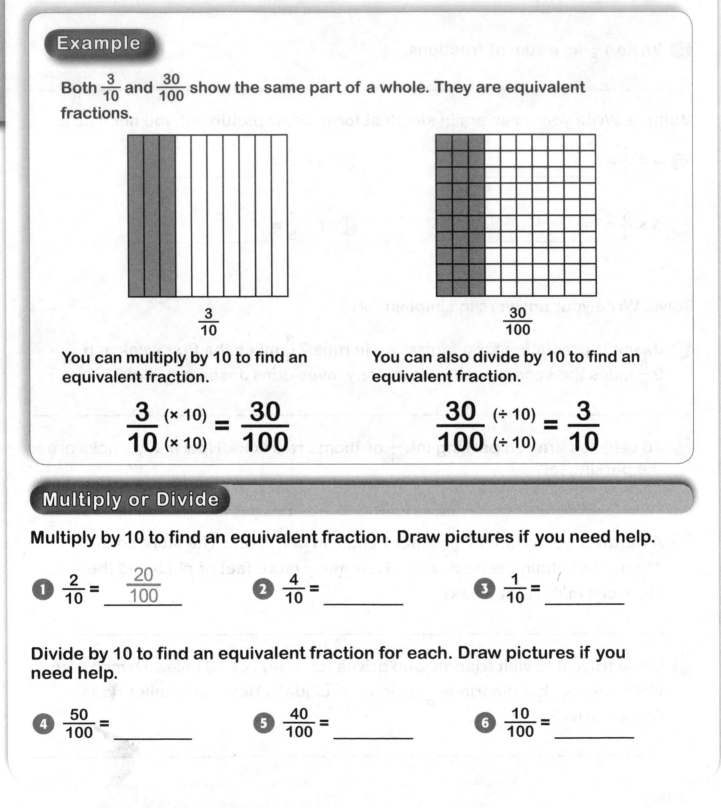

Example

Both $\frac{3}{10}$ and $\frac{30}{100}$ show the same part of a whole. They are equivalent fractions.

$\frac{3}{10}$

$\frac{30}{100}$

You can multiply by 10 to find an equivalent fraction.

You can also divide by 10 to find an equivalent fraction.

$$\frac{3 \ (\times 10)}{10 \ (\times 10)} = \frac{30}{100}$$

$$\frac{30 \ (\div 10)}{100 \ (\div 10)} = \frac{3}{10}$$

Multiply or Divide

Multiply by 10 to find an equivalent fraction. Draw pictures if you need help.

1. $\frac{2}{10} = \frac{20}{100}$

2. $\frac{4}{10} =$ _____

3. $\frac{1}{10} =$ _____

Divide by 10 to find an equivalent fraction for each. Draw pictures if you need help.

4. $\frac{50}{100} =$ _____

5. $\frac{40}{100} =$ _____

6. $\frac{10}{100} =$ _____

Adding Fractions with Denominators 10 and 100

Examples

Find $\frac{4}{10} + \frac{3}{10}$.

You can add the numerators because the denominators are the same.

$$\frac{4}{10} + \frac{3}{10} = \frac{4+3}{10} = \frac{7}{10}$$

Find $\frac{2}{10} + \frac{72}{100}$.

The denominators are different. Convert the fraction with denominator 10 to an equivalent fraction with denominator 100. Then add.

$$\frac{2}{10} \, {}^{(\times 10)}_{(\times 10)} = \frac{20}{100}$$

$$\frac{20}{100} + \frac{72}{100} = \frac{20+72}{100} = \frac{92}{100}$$

Add

1 $\frac{3}{10} + \frac{3}{10} = \underline{\frac{6}{10}}$

2 $\frac{50}{100} + \frac{20}{100} = \underline{\hspace{2cm}}$

3 $\frac{6}{10} + \frac{2}{10} = \underline{\hspace{2cm}}$

4 $\frac{4}{10} + \frac{1}{10} = \underline{\hspace{2cm}}$

5 $\frac{21}{100} + \frac{47}{100} = \underline{\hspace{2cm}}$

6 $\frac{7}{10} + \frac{3}{10} = \underline{\hspace{2cm}}$

7 Draw three squares of equal size. Show $\frac{8}{10}$ on one square, $\frac{80}{100}$ on another, and $\frac{4}{5}$ on the third. Explain what you see.

_____

Name _____

Fractions as Decimals

You can write fractions or mixed numbers with denominators of 10 or 100 as decimals.

Example

A decimal is a number with one or more digits to the right of a decimal point.

3.42

Decimal point

You can use drawings to show decimals.

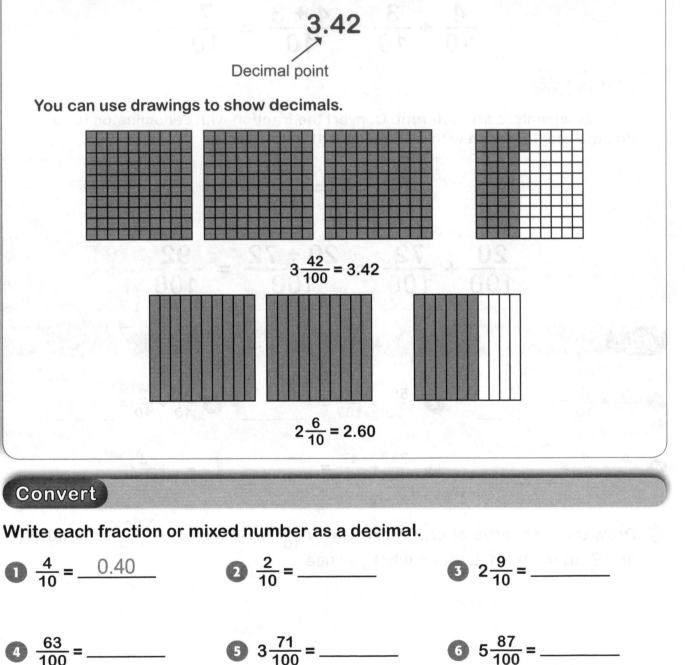

$$3\frac{42}{100} = 3.42$$

$$2\frac{6}{10} = 2.60$$

Convert

Write each fraction or mixed number as a decimal.

1. $\frac{4}{10}$ = ___0.40___

2. $\frac{2}{10}$ = _____

3. $2\frac{9}{10}$ = _____

4. $\frac{63}{100}$ = _____

5. $3\frac{71}{100}$ = _____

6. $5\frac{87}{100}$ = _____

Decimals and Fractions on a Number Line

You can write fractions and decimals on a number line.

Examples

Write $\frac{1}{4}$ on a number line.

Draw a number line. Label 0 and 1. Divide the distance between 0 and 1 into 4 equal parts. Draw a point at $\frac{1}{4}$.

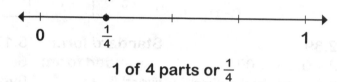

1 of 4 parts or $\frac{1}{4}$

Write 0.60 on a number line.

Draw a number line. Label 0 and 1. Divide the distance between 0 and 1 into 10 equal parts. Draw a point at 0.60.

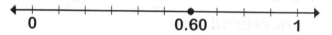

Write 0.38 on a number line.

Draw a number line. Label 0 and 1. Divide the distance between 0 and 1 into 10 equal parts. Draw a point at the approximate location of 0.38.

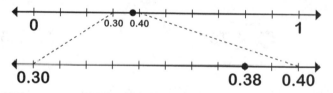

There are 10 equal parts between 0.30 and 0.40. Each of these parts is 0.01.

Plot

Plot each decimal point or fraction on the number line provided.

1. Plot $\frac{1}{2}$ on the number line. Label it A.

2. Plot 0.30 on the number line. Label it B.

3. Plot $\frac{8}{10}$ on the number line. Label it C.

Name _____

Place Value of Decimals

Digits in decimal numbers have place values. You can show them in standard form, expanded form, and word form.

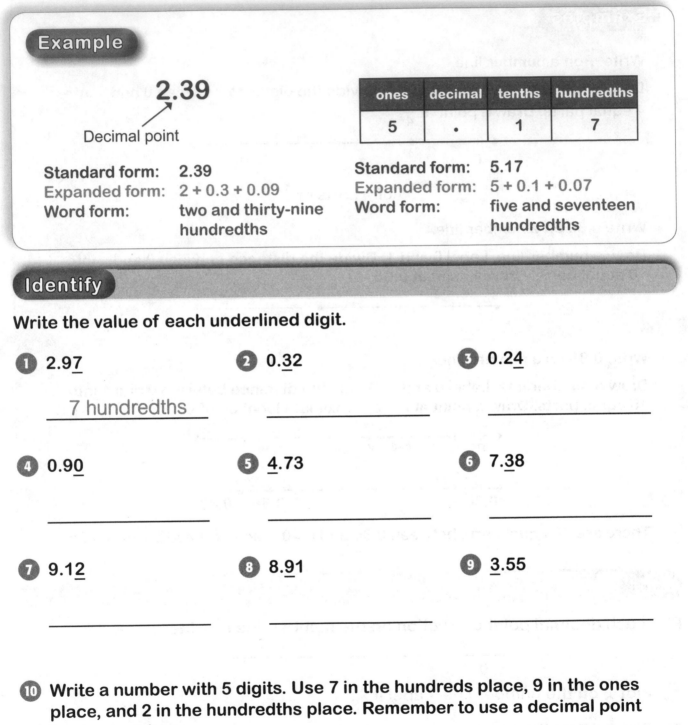

Example

2.39

Decimal point

ones	decimal	tenths	hundredths
5	.	1	7

Standard form: 2.39
Expanded form: 2 + 0.3 + 0.09
Word form: two and thirty-nine hundredths

Standard form: 5.17
Expanded form: 5 + 0.1 + 0.07
Word form: five and seventeen hundredths

Identify

Write the value of each underlined digit.

1 2.9<u>7</u>

____7 hundredths____

2 0.<u>3</u>2

_____

3 0.2<u>4</u>

_____

4 0.9<u>0</u>

_____

5 <u>4</u>.73

_____

6 7.<u>3</u>8

_____

7 9.1<u>2</u>

_____

8 8.<u>9</u>1

_____

9 <u>3</u>.55

_____

10 Write a number with 5 digits. Use 7 in the hundreds place, 9 in the ones place, and 2 in the hundredths place. Remember to use a decimal point in your number.

_____

Comparing Decimals

You can compare decimal numbers using place values. You can use the symbols <, >, and = to tell how they compare.

Example

Which number is greater, 3.04 or 3.39?

Step 1: Write the numbers. Line up the digits. Start at the left and compare.

$$\underline{3}.14$$
$$\underline{3}.39$$

The ones digit is the same in both numbers.

Step 2: Look at the next digit. Keep looking until you come to digits that are different.

$$3.\underline{1}4$$
$$3.\underline{3}9$$

The tens digits are different.

Step 3: Compare the digits that are different.

$$3.\underline{1}4$$
$$3.\underline{3}9$$

1 tenth is less than 3 tenths.
So 3.14 is less than 3.39.

Remember! The symbol > means greater than. The symbol < means less than.

$$3.14 < 3.39$$

Compare

Compare each pair of decimal numbers. Write <, >, or = for each.

1 0.04 __<__ 0.06

2 0.37 _____ 0.23

3 0.55 _____ 0.52

4 6.12 _____ 6.19

5 8.08 _____ 8.08

6 5.3 _____ 5.23

7 A blue jay has a wingspan of 14.5 inches. A robin has a wingspan of 14.42 inches. Which bird has the greater wingspan?

_____

Name _____

Valid Comparisons

When you compare decimal numbers, you must be sure they refer to the same whole. Sometimes drawings can help.

Examples

Mr. Taylor plants tomatoes in 0.25 of his garden. He plants green beans in 0.50 of his garden. He says he planted more green beans than tomatoes. Is he correct?

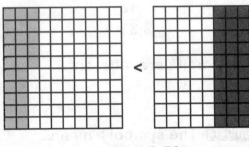

0.25 < 0.50

He is correct because the comparison refers to the same whole. Both decimal numbers show parts of the same garden.

Mr. Torres has a board that is 0.25 of a foot long. His son has a board that is 0.25 of a yard long. Are the boards the same length? How do you know?

```
                    1 yard
  ┌──────┬──────────────────────────────┐
  │ 0.25 │                              │
  └──────┴──────────────────────────────┘

   0.25        1 foot
        ┌──┬────────────┐
        │  │            │
        └──┴────────────┘
```

The boards are not the same length. The decimal numbers are equal, but they do not refer to the same whole.

Solve

1 Greg hiked a trail that is 3.48 miles long. George hiked a trail that is 2.97 miles long. Greg says he hiked farther than George. Is he correct? Why or why not?

_____

_____

2 Karen studied 0.85 hours for a test. Seina studied 0.51 days for the same test. Who studied longer? How do you know?

_____

_____

Decimals and Money

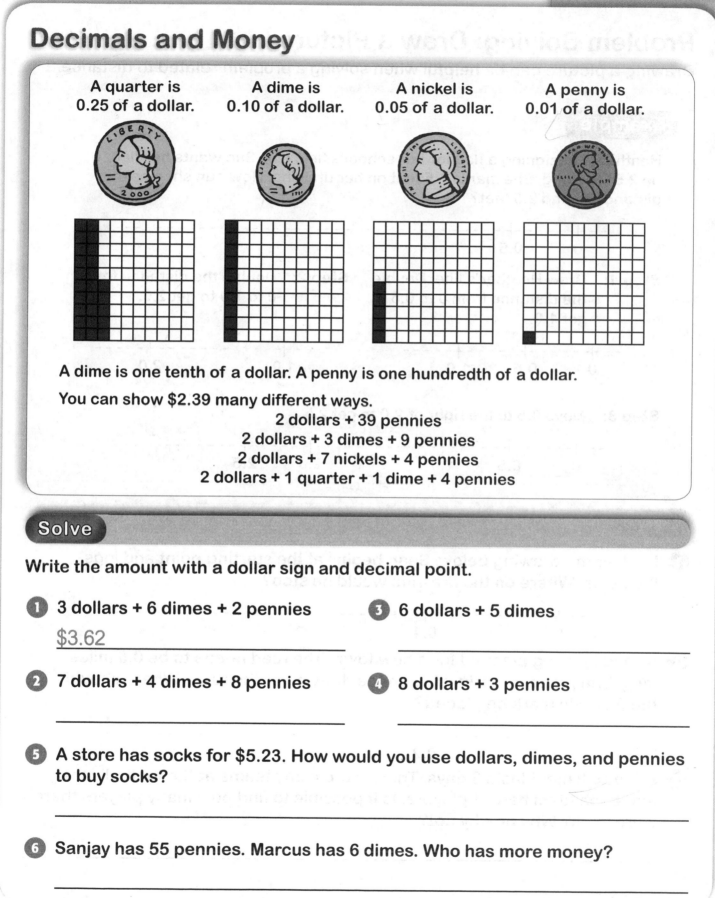

| A quarter is 0.25 of a dollar. | A dime is 0.10 of a dollar. | A nickel is 0.05 of a dollar. | A penny is 0.01 of a dollar. |

A dime is one tenth of a dollar. A penny is one hundredth of a dollar.

You can show $2.39 many different ways.

2 dollars + 39 pennies
2 dollars + 3 dimes + 9 pennies
2 dollars + 7 nickels + 4 pennies
2 dollars + 1 quarter + 1 dime + 4 pennies

Solve

Write the amount with a dollar sign and decimal point.

1 3 dollars + 6 dimes + 2 pennies

$3.62 _____

3 6 dollars + 5 dimes

_____

2 7 dollars + 4 dimes + 8 pennies

_____

4 8 dollars + 3 pennies

_____

5 A store has socks for $5.23. How would you use dollars, dimes, and pennies to buy socks?

_____

6 Sanjay has 55 pennies. Marcus has 6 dimes. Who has more money?

_____

Name _____

Problem Solving: Draw a Picture

Drawing a picture can be helpful when solving a problem related to distance.

Example

Heather is designing a flag for her school's field day. She wants her flag to be 2.5 feet long. She marks 0.5 feet on her drawing. How can she use this distance to find 2.5 feet?

```
- - -|- - - - - -|- - - - - - - - - - - - - - - - - - - - - - - - - -
    0       0.5
```

Step 1: Draw Heather's line. Double the distance from 0 to 0.5 to get 1.0.

```
- - -|- - - - - -|- - - - -|- - - - -
    0       0.5      1.0
```

Step 2: Double the distance from 0 to 1.0 to get 2.0.

```
- - -|- - - - - - - - - - - -|- - - - -
    1.0                 2.0
```

Step 3: Move 0.5 to the right of 2.0 to get 2.5.

```
- - -|- - - - - -|- - - - -|- - - - - - - - - - -|- - - - -|- - - - -
    0       0.5      1.0                2.0     2.5
```

Solve

1 Look at the drawing below. Sean begins at the starting point and jogs 0.8 miles. Where on the drawing would he stop?

```
- - -|- - - - - - - - - -|- - - - - - - - -|- - - - - - - - - - - - -
    0                 0.4               0.8
```

2 A road is being planned for a new town. The road needs to be 0.5 miles long. The planner marked 0.2 on the drawing of the road. Where should the 0.5-mile mark be placed?

```
- - -|- - - - - - - - - -|- - - - - - - - - - - - - - - - - - - - - -
    0                 0.2
```

3 A football meet lasts 3 days. There were many teams at the football meet, and each team had 11 players. Is it possible to find how many players there were in all? Why or why not?

_____

Name _____

Multiply by 10 to find an equivalent fraction for each.

1 $\frac{5}{10}$ _____

2 $\frac{3}{10}$ _____

3 $\frac{8}{10}$ _____

Divide by 10 to find an equivalent fraction for each.

4 $\frac{20}{100}$ _____

5 $\frac{10}{100}$ _____

6 $\frac{70}{100}$ _____

Add. Write your answers in simplest form.

7 $\frac{1}{10} + \frac{3}{10} =$ _____

8 $\frac{46}{100} + \frac{25}{100} =$ _____

9 $\frac{4}{10} + \frac{17}{100} =$ _____

Plot each decimal point or fraction on the number line provided.

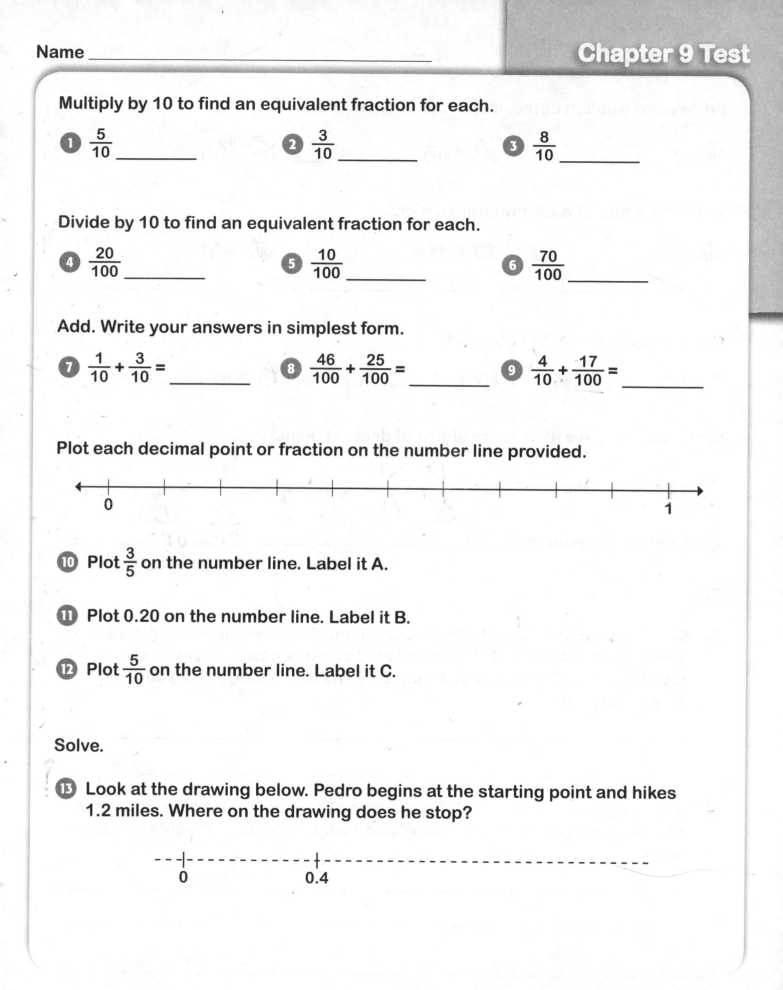

10 Plot $\frac{3}{5}$ on the number line. Label it A.

11 Plot 0.20 on the number line. Label it B.

12 Plot $\frac{5}{10}$ on the number line. Label it C.

Solve.

13 Look at the drawing below. Pedro begins at the starting point and hikes 1.2 miles. Where on the drawing does he stop?

Write each fraction or mixed number as a decimal.

14 $\frac{6}{10}$ _____

15 $2\frac{78}{100}$ _____

16 $42\frac{7}{10}$ _____

Write the value of each underlined digit.

17 1.2<u>5</u>

18 0.<u>4</u>8

19 <u>3</u>.56

_____ _____ _____

Compare. Write <, >, or = for each.

20 0.41 _____ 0.14

21 1.72 _____ 1.72

22 3.81 _____ 3.85

Write the amount with a dollar sign and decimal point.

23 1 dollar + 4 dimes + 7 pennies _____

24 2 dollars + 5 pennies _____

Solve.

25 Benjamin and Bob went to the beach. Benjamin drew a line in the sand. It measured 3.48 feet. Bob drew a line that measured 2.39 yards. Benjamin says that since 3.48 is greater than 2.39, his line is longer. Do you agree? Why or why not?

_____

_____

26 A German shepherd named Mindy weighs 74.09 pounds. Another German shepherd weighs 74.32 pounds. Would comparing these weights be valid? Why?

_____

_____

Name _____

Metric Units of Length

The meter, the centimeter, and the kilometer are metric units of length. You can convert measurements in one metric unit to another metric unit.

Examples

A bean is about 1 centimeter long.

A doorknob is about 1 meter above the floor.

Five city blocks is about 1 kilometer.

You can convert measurements in one unit to another unit.

1 kilometer (km) = 1,000 meters (m)
1 meter (m) = 100 centimeters (cm)

This school bus is about 10 meters long.
10 × 100 centimeters = 1,000 centimeters
This school bus is about 1,000 centimeters long.

Solve

Place a check mark next to the best answer.

1 About how long is a bee?

____ about 1 meter

✓ about 1 centimeter

____ about 1 kilometer

3 About how long is a new pencil?

____ about 16 meters

____ about 16 centimeters

____ about 16 kilometers

2 About how long is a car?

____ about 3 meters

____ about 3 centimeters

____ about 3 kilometers

4 About how long is a city block?

____ about 1 kilometer

____ about 11 meters

____ about 111 centimeters

Multiply to find each missing number.

5 30 km = _____ m   **6** 5 m = _____ cm   **7** 8 km = _____ m

Using Tables with Metric Units of Length

You can draw tables to record measurements and show conversions.

Example

Ms. Roy has measurements of the lengths of different vehicles to the nearest meter. She wants to convert these measurements to centimeters. How can she show this information easily?

Mr. Desai's car: 5 meters

Ms. Brown's car: 4 meters

Mr. Vega's truck: 6 meters

Step 1: Draw a table to show the measurements.

Step 2: Multiply to convert each measurement. Complete the table.

	Meters	Centimeters
Mr. D's car	5	
Ms. B's car	4	
Mr. V's truck	6	

	Meters	Centimeters
Mr. D's car	5	500
Ms. B's car	4	400
Mr. V's truck	6	600

Solve

1. Draw a table to show the following data. Then convert each length to meters.

   Chase Street: 3 kilometers

   Maple Street: 4 kilometers

   Lower Bridge Street: 7 kilometers

   Green Avenue: 8 kilometers

2. Draw a table to show the following data. Then convert each length to centimeters.

   Height of Corn Plant: 2 meters

   Height of Sunflower Plant: 3 meters

   Height of Young Tree: 5 meters

   Height of Old Tree: 18 meters

Name _____

Customary Units of Length

The mile, the yard, the foot, and the inch are customary units of length. You can convert measurements in one customary unit to another customary unit.

A paperclip is about 1 inch long.

A loaf of bread is about 1 foot long.

A baseball bat is about 1 yard long.

Most people can jog a mile in about 10 minutes.

You can convert measurements in one unit to another unit.
1 mile (mi) = 1,760 yards (yd)   1 mile (mi) = 5,280 feet (ft)
1 yard (yd) = 36 inches (in.)   1 yard (yd) = 3 feet (ft)
1 foot (ft) = 12 inches (in.)

The bench is about 5 feet long.
5 × 12 inches = 60 inches
The bench is about 60 inches long.

Solve

Place a check mark next to the best answer.

1 About how tall is a stamp?
____ about 1 yard
____ about 1 foot
✓ about 1 inch

2 About how long is a kite?
____ about 1 yard
____ about 1 foot
____ about 1 inch

3 About how long is a car?
____ about 14 yards
____ about 14 feet
____ about 14 inches

4 About how long is a golf course?
____ about 5 feet
____ about 5 yards
____ about about 5 miles

Multiply to find each missing number.

5 8 ft = _____ in.   **6** 7 yd = _____ ft   **7** 4 mi = _____ yd

8 13 yd = _____ in.   **9** 75 ft = _____ in.   **10** 6 mi = _____ ft

Lesson 4

Name _____

Using Tables with Customary Units of Length
You can draw tables to help you convert measurements.

Example

Mr. Neal is building a doghouse. He wants to convert a list of measurements to help him decide how to saw the wood. How can he show this information easily?

Walls: 3 ft
Floor: 5 ft
Roof: 4 ft
Door: 2 ft

Step 1: Draw a table to show the measurements.

Step 2: Multiply to convert each measurement. Complete the table.

	Feet	Inches
Walls	3	
Floor	5	
Roof	4	
Door	2	

	Feet	Inches
Walls	3	36
Floor	5	60
Roof	4	48
Door	2	24

Solve

1. Draw a table to show the following data. Then convert each length to yards.

   Store to Park: 1 mile

   Store to Home: 2 miles

   Store to Library: 3 miles

   Store to Bank: 4 miles

2. Complete the table below.

yd	1	2	3	4	5	6
ft						
in.						

Finding Area

You can use the formula $A = l \times w$ to find the area of squares and rectangles.

Example

Alex wants to tile a floor in his house. The floor measures 14 feet by 8 feet. He has 1-foot square tiles. How many tiles will Alex use?

When Alex places the tiles, they form a grid similar to an array. You can count the tiles or multiply to find the number of tiles in all.

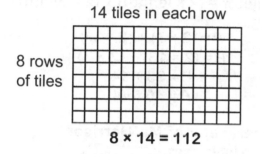

14 tiles in each row

8 rows of tiles

8 × 14 = 112

The number of square units needed to cover a region is called area. You can find the area of squares and rectangles using a formula: Area = length × width.

$$A = l \times w$$
$$A = 8 \text{ feet} \times 14 \text{ feet}$$
$$A = 112 \text{ square feet}$$

The area of Alex's floor is 112 square feet.

Alex will use 112 tiles.

Area is always measured in square units, such as square feet, square meters, or square miles.

Calculate

Find the area of each figure.

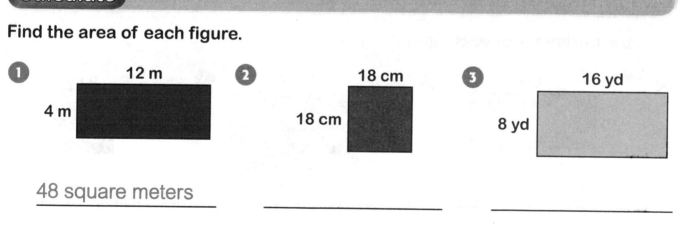

1 12 m 4 m

2 18 cm 18 cm

3 16 yd 8 yd

48 square meters

_____   _____

4 Hunter wants to paint a wall in a large office. The wall measures 7 meters by 9 meters. How much area does Hunter need to paint?

_____

Name _____

Finding Perimeter

You can use the formula $P = (2 \times l) + (2 \times w)$ to find the perimeter of squares and rectangles.

Example

Ms. Harrison wants to put a fence around her garden. Her garden is 13 feet long and 7 feet wide. How much fence will Ms. Harrison need?

Perimeter is the distance around a figure. To find perimeter, you can measure the length of each side. Then add the lengths.

13 feet long

7 feet wide

$$13 + 13 + 7 + 7 = 40$$

Ms. Harrison needs 40 feet of fence.

You can find the perimeter of squares and rectangles using a formula:
Perimeter = (2 × length) + (2 × width).

$$P = (2 \times l) + (2 \times w)$$
$$P = (2 \times 13) + (2 \times 7)$$
$$P = 26 + 14$$
$$P = 40 \text{ feet}$$

The perimeter of Ms. Harrison's garden is 40 feet.

Calculate

Find the perimeter of each figure.

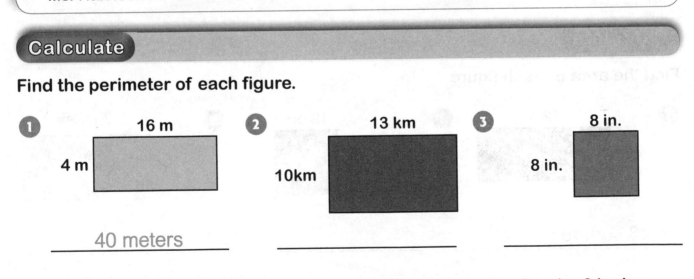

1   16 m   4 m

_____40 meters_____

2   13 km   10km

_____

3   8 in.   8 in.

_____

4 Sophia is making a frame for a photo. If the photo is 4 inches by 6 inches, how much wood will she need for the frame?

_____

Problem Solving: Draw a Picture and Write an Equation

Drawing a picture and writing an equation can help you solve problems.

Example

Josh's cat is $68\frac{3}{4}$ cm long from nose to tail. Marisa's cat is $64\frac{1}{4}$ cm long from nose to tail. How much longer is Josh's cat?

Step 1: Read the problem. Do you need to add, subtract, multiply, or divide? Draw a picture and write an equation to show the problem.

$$68\frac{3}{4}$$

$64\frac{1}{4}$	?

$$68\frac{3}{4} - 64\frac{1}{4} = ?$$

Step 2: Subtract to solve. Simplify your answer if necessary. Remember to include the unit of measurement in your answer.

$$\begin{array}{r} 68\frac{3}{4} \\ - 64\frac{1}{4} \\ \hline 4\frac{2}{4} = 4\frac{1}{2} \end{array}$$

Josh's cat is $4\frac{1}{2}$ cm longer than Marisa's.

Solve

1 Elsa is knitting a blanket. She used $4\frac{3}{8}$ meters of red yarn and $2\frac{4}{8}$ meters of gold yarn. How many meters of yarn did she use altogether?

$6\frac{7}{8}$ meters

2 Katia is building a tree house. She needs 52 boards. Each board has to be 4 meters long. How much wood does Katia need?

_____

3 Ms. Morton has $167\frac{3}{5}$ cm of string. She cuts the string into two pieces. One piece is $72\frac{1}{5}$ cm long. How long is the other piece?

_____

Name _____

Problem Solving: Working Backward

Some problems are easier to solve if you "work backward." Operation trains can help you do this.

Example

Franklin has a spool of string. He gave $22\frac{1}{8}$ inches to his sister Gina. He also gave $49\frac{2}{8}$ inches to a friend. He has 56 inches left. How many inches of string did Franklin start with?

Step 1: Write an operation train to show the problem. Use variables to show unknown numbers.

$$x - 22\frac{1}{8} - 49\frac{2}{8} = 56$$

Step 2: Now you know the end number and how it was calculated. You can work backward using inverse operations to find the beginning number.

$$56 + 49\frac{2}{8} + 22\frac{1}{8} = x$$

$$x = 127\frac{3}{8}$$

Franklin started with $127\frac{3}{8}$ inches of string.

Solve

1. The owner of a gift stop has a large spool of wrapping paper. She uses $13\frac{2}{5}$ yards of paper in one week and $16\frac{1}{5}$ yards of paper the next week. She has $41\frac{1}{5}$ yards of paper left over. How many yards of wrapping paper did she start with?

_____

2. Toshi drove from home to work. Then he drove from work to Beth's house. He drove $13\frac{1}{4}$ miles south, then 29 miles east, then another $8\frac{1}{4}$ miles south. He drove a total of $55\frac{3}{4}$ miles for the entire day. How many miles did he drive from home to work?

_____

Problem Solving: Using Formulas

You can use formulas to solve problems. You have to read the problem carefully to know which formula to use.

Example

A rectangular living room measures 19 feet long by 9 feet wide. Ms. Perez wants to put new tile on the floor. How many square feet of tile will she need?

Step 1:
Read the problem carefully. You know the living room is rectangular. It measures 19 feet long and 9 feet wide. You are asked to find how many square feet of tile Ms. Perez will need. Can you use a formula to solve the problem?

Step 2:
You can use $A = l \times w$ to find the total amount of carpet. Apply the formula and solve.

$$A = l \times w$$
$$A = 19 \times 9$$
$$A = 171 \text{ square feet}$$

Ms. Perez will need 171 square feet of tile for the living room.

Remember! The formula for finding perimeter is $P = (2 \times l) + (2 \times w)$.

Solve

1 Mr. Smith wants to build a fence around his backyard. It is square-shaped and measures 7 meters on one side. How many meters of fence will Mr. Smith need to build?

$(2 \times 7) + (2 \times 7) = 28m$

2 Gary builds a rectangular rabbit pen. The pen measures 6 feet by 8 feet. How many square feet is the pen?

_____

3 The town of Grand Sun marks off a large rectangular patch of land as a forest preserve. The land measures 14 kilometers long and 8 kilometers wide. What is the total area of the forest preserve?

_____

Name _____

Place a check mark next to the best answer.

1 About how long is a paperclip?

____ about 1 inch

____ about 1 foot

____ about 1 yard

2 About how long is a horse?

____ about 2 centimeters

____ about 2 kilometers

____ about 2 meters

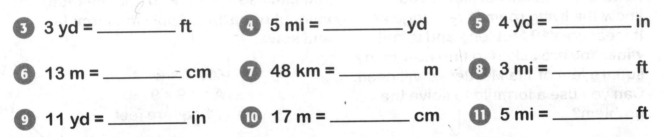

Multiply to find each missing number.

3 3 yd = _____ ft

4 5 mi = _____ yd

5 4 yd = _____ in

6 13 m = _____ cm

7 48 km = _____ m

8 3 mi = _____ ft

9 11 yd = _____ in

10 17 m = _____ cm

11 5 mi = _____ ft

12 Draw a table to show the following data.
Then convert each length to inches.
Show both lengths in the table.

Maple Tree: 32 feet

Oak Tree: 75 feet

Birch Tree: 56 feet

Pine Tree: 95 feet

13 Draw a table to show the following data.
Then convert each length to centimeters.
Show both lengths in the table.

Height of Garden Shed: 3 meters

Height of Garage: 4 meters

Height of House: 22 meters

Height of Mailbox: 1 meter

14 What is the area of the rectangle?

15 in.

5 in.

_____

15 What is the area of the square?

13 m

13 m

_____

16 What is the perimeter of the rectangle?

18 km

9 km

_____

Solve.

17 Lauren and her brother went hiking. They hiked $44\frac{3}{8}$ kilometers along the Atlanta Trail. They also hiked $36\frac{2}{8}$ kilometers along the Hopper Trail. How many kilometers did they hike in all?

_____

18 A fabric store has a sale. $22\frac{3}{5}$ yards of black fabric are sold the first day and $19\frac{1}{5}$ yards are sold the second day. $16\frac{1}{5}$ yards of black fabric are left over. How many yards of black fabric were in the store to begin with?

_____

19 Mr. Harrison is an architect. He designs a rectangular living room that is 8 meters wide and 11 meters long. What is the area of the living room?

_____

20 What is the perimeter of the living room?

_____

21 Karen has a new roll of tape. She uses $3\frac{1}{6}$ feet to tape one package. She uses $5\frac{2}{6}$ feet to tape another package. Then she uses $4\frac{1}{6}$ feet to tape a third package. She has $12\frac{2}{6}$ feet of tape left over. How many feet of tape did she begin with?

_____

Name _____

Metric Units of Mass

The kilogram and the gram are metric units of mass. You can convert kilograms to grams easily.

Example

Weight is a measure of how heavy something is. Weight changes depending on location. The weight of a pumpkin on the moon is not the same as its weight on Earth.

Mass is a measure of the amount of matter in an object. Mass never changes. The mass of the pumpkin is the same on the moon and on Earth.

A feather has a mass of about 1 gram.
A pineapple has a mass of about 1 kilogram.

You can convert measurements from kilograms to grams.

1 kilogram (kg) = 1,000 grams (g)

A bowling ball is about 6 kilograms.
6 × 1,000 grams = 6,000 grams
A bowling ball is about 6,000 grams.

Solve

Place a check mark beside the best answer.

1 What is the mass of a paperclip?

✓ about 1 gram   ____ about 1 kilogram

____ about 50 grams   ____ about 100 grams

2 What is the mass of a cat?

____ about 3 grams   ____ about 300 kilograms

____ about 30 grams   ____ about 3 kilograms

Multiply to find each missing number.

3 1 kg = _____ g   **4** 4 kg = _____ g   **5** 13 kg = _____ g

6 57 kg = _____ g   **7** 472 kg = _____ g   **8** 397 kg = _____ g

Customary Units of Weight

The ounce, the pound, and the ton are customary units of weight. You can convert measurements from one unit to another.

Examples

A slice of bread weighs about 1 ounce.

A bunch of grapes weighs about 1 pound.

A small car weighs about 1 ton.

You can convert measurements from one unit to another unit.

1 ton (T) = 2,000 pounds (lb)
1 pound (lb) = 16 ounces (oz)

A bicycle is about 30 pounds.
30 × 16 ounces = 480 ounces
The bicycle is about 480 ounces.

Solve

Place a check mark next to the best answer.

1 What does a baby rabbit weigh?

____ about 1 ounce   ✓ about 1 pound   ____ about 10 pounds

2 What does a cherry weigh?

____ about 1 ounce   ____ about 10 ounces   ____ about 1 pound

3 What does a gallon jug of milk weigh?

____ about 2 pounds   ____ about 9 pounds   ____ about 25 pounds

4 Multiply to complete the tables.

Tons	1	2	3	4	5	6
Pounds	2,000					

Pounds	1	2	3	4	5	6
Ounces	16					

Metric Units of Capacity

The milliliter and the liter are metric units of capacity, or volume. You can convert liters to milliliters easily.

Examples

The volume of a container measured in liquid units is called capacity.

An eyedropper will hold about 1 milliliter.

Many water bottles hold about 1 liter.

You can convert measurements from liters to milliliters.
1 liter (L) = 1,000 milliliters (mL)

A watering can holds about 2 liters.
2 × 1,000 milliliters = 2,000 milliliters
The watering can holds about 2,000 milliliters.

Solve

Place a check mark next to the best answer.

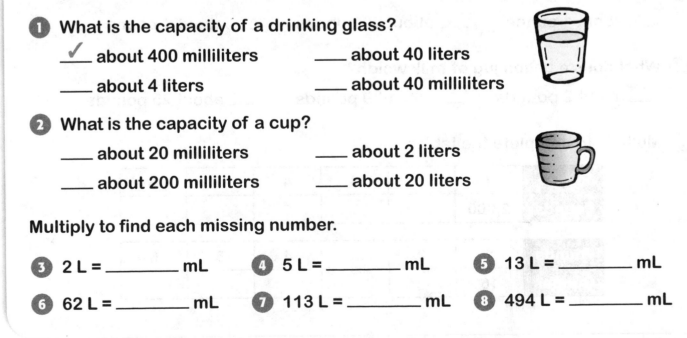

1 What is the capacity of a drinking glass?

✓ about 400 milliliters   ____ about 40 liters

____ about 4 liters   ____ about 40 milliliters

2 What is the capacity of a cup?

____ about 20 milliliters   ____ about 2 liters

____ about 200 milliliters   ____ about 20 liters

Multiply to find each missing number.

3 2 L = _____ mL

4 5 L = _____ mL

5 13 L = _____ mL

6 62 L = _____ mL

7 113 L = _____ mL

8 494 L = _____ mL

Customary Units of Capacity

The cup, the pint, the quart, and the gallon are customary units of capacity, or volume. You can convert measurements from one unit to another.

Examples

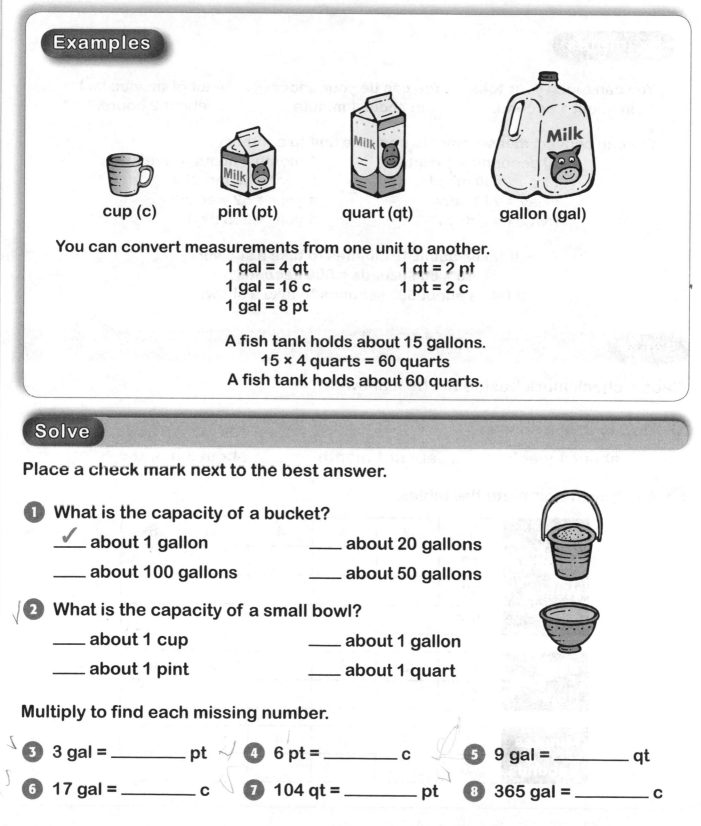

cup (c)   pint (pt)   quart (qt)   gallon (gal)

You can convert measurements from one unit to another.

1 gal = 4 qt	1 qt = 2 pt
1 gal = 16 c	1 pt = 2 c
1 gal = 8 pt	

A fish tank holds about 15 gallons.
15 × 4 quarts = 60 quarts
A fish tank holds about 60 quarts.

Solve

Place a check mark next to the best answer.

1 What is the capacity of a bucket?

✓ about 1 gallon   ____ about 20 gallons

____ about 100 gallons   ____ about 50 gallons

2 What is the capacity of a small bowl?

____ about 1 cup   ____ about 1 gallon

____ about 1 pint   ____ about 1 quart

Multiply to find each missing number.

3 3 gal = _____ pt   **4** 6 pt = _____ c   **5** 9 gal = _____ qt

6 17 gal = _____ c   **7** 104 qt = _____ pt   **8** 365 gal = _____ c

Name _____

Units of Time

The second, the minute, the hour, the day, the week, the month, and the year are units of time. You can convert units from one to another.

Examples

You can touch your toes in about 1 second.

You can tie your shoes in about 1 minute.

A lot of movies last about 2 hours.

You can convert measurements from one unit to another.

60 seconds = 1 minute	1 month = about 4 weeks
1 hour = 60 minutes	1 year = 12 months
1 day = 24 hours	1 year = 52 weeks
1 week = 7 days	1 year = 365 days

It takes about 10 minutes to take a shower.
10×60 seconds = 600 seconds
It takes about 600 seconds to take a shower.

Solve

Place a check mark beside the best answer.

1 How long is a season?

____ about 1 week   ____ about 1 month   ____ about 3 months

2 Multiply to complete the tables.

Years	1	2	3	4	5	6
Months	12					
Weeks	52					
Days	365					

Days	1	2	3	4	5	6
Hours	24					

Minutes	1	2	3	4	5	6
Seconds	60					

Using Line Plots

Line plots use number lines to show data.

Example

This table shows the height of seven students.

Name	Jake	Lisa	Quentin	Blake	Akemi	Farah	Pedro
Height in Meters	$1\frac{2}{10}$	$1\frac{4}{10}$	$1\frac{3}{10}$	$1\frac{7}{10}$	$1\frac{2}{10}$	$1\frac{3}{10}$	$1\frac{3}{10}$

A line plot shows data. Each X shows one number from the data set.

The most Xs are above $1\frac{3}{10}$. That is the most common height in meters.

The greatest height shown is $1\frac{7}{10}$. The shortest height shown is $1\frac{2}{10}$.

$1\frac{7}{10}$ is far from the other numbers in the data set. It is an **outlier**. An outlier is any number that is very different from the rest of the numbers in the set.

Interpret and Diagram

The line plot shows the mass in kilograms of eight dogs.

❶ What is the most common mass?

   $32\frac{1}{2}$ kg
   _____

❷ What is the greatest mass shown on the line plot?

   _____

❸ Which mass is an outlier?

   _____

❹ How many dogs have a mass of 33 kg?

   _____

Using Line Plots to Solve Problems

Sometimes you will need to read line plots to solve problems.

Example

Students from a science class collect rocks for a science exhibit. This line plot shows the mass of different rocks gathered by the class.

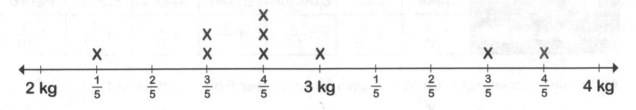

Will's rock is shown with the red X. Holly's rock is shown with the blue X. What is the difference in mass between Holly's rock and Will's rock?

Step 1: Read the line plot to find the information you need.

the mass of Will's rock = $2\frac{3}{5}$ kg

the mass of Holly's rock = $3\frac{4}{5}$ kg

Step 2: Add or subtract to solve. To find the difference, you subtract.

$$3\frac{4}{5} - 2\frac{3}{5} = 1\frac{1}{5}$$

The difference in mass between Holly's rock and Will's rock is $1\frac{1}{5}$ kg.

Solve

The line plot shows the height in feet of a number of parade balloons.

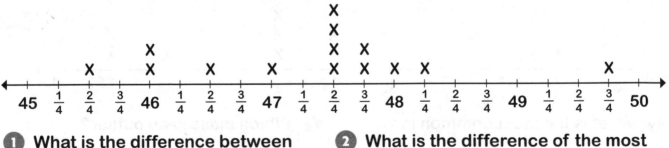

1 What is the difference between the tallest balloon and the shortest balloon shown on the line plot?

_____

2 What is the difference of the most common height shown and the shortest height shown?

_____

Problem Solving: Using a Number Line

You can add and subtract measurements using a number line.

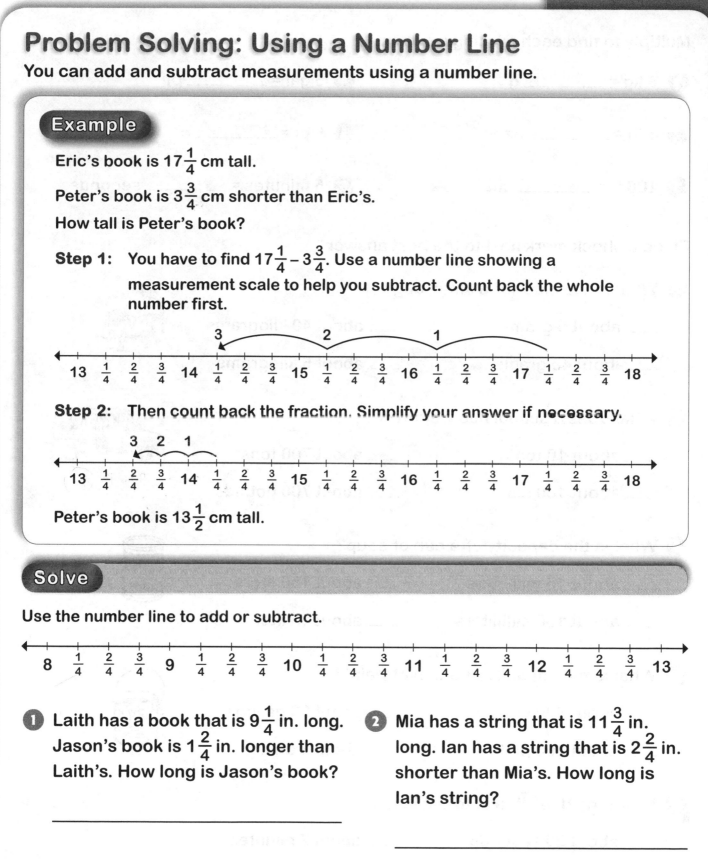

Example

Eric's book is $17\frac{1}{4}$ cm tall.

Peter's book is $3\frac{3}{4}$ cm shorter than Eric's.

How tall is Peter's book?

Step 1: You have to find $17\frac{1}{4} - 3\frac{3}{4}$. Use a number line showing a measurement scale to help you subtract. Count back the whole number first.

Step 2: Then count back the fraction. Simplify your answer if necessary.

Peter's book is $13\frac{1}{2}$ cm tall.

Solve

Use the number line to add or subtract.

1. Laith has a book that is $9\frac{1}{4}$ in. long. Jason's book is $1\frac{2}{4}$ in. longer than Laith's. How long is Jason's book?

_____

2. Mia has a string that is $11\frac{3}{4}$ in. long. Ian has a string that is $2\frac{2}{4}$ in. shorter than Mia's. How long is Ian's string?

_____

Name _____

Multiply to find each missing number.

1 3 kg = _____ g

4 5 gal = _____ pt

2 3 lb = _____ oz

5 5 pt = _____ c

3 106 L = _____ mL

6 5 minutes = _____ seconds

Place a check mark next to the best answer.

7 What is the mass of a small dog?

____ about 5 grams ____ about 40 kilograms

____ about 40 grams ____ about 5 kilograms

8 What does a school bus weigh?

____ about 10 tons ____ about 700 tons

____ about 100 lbs ____ about 700 pounds

9 What is the capacity of a can of soup?

____ about 15 milliliters ____ about 150 liters

____ about 150 milliliters ____ about 15 liters

10 What is the capacity of a can of paint?

____ about 300 cups ____ about 30 gallons

____ about 300 pints ____ about 3 quarts

11 How long does it take to eat lunch?

____ about 20 seconds ____ about 2 minutes

____ about 20 minutes ____ about 2 hours

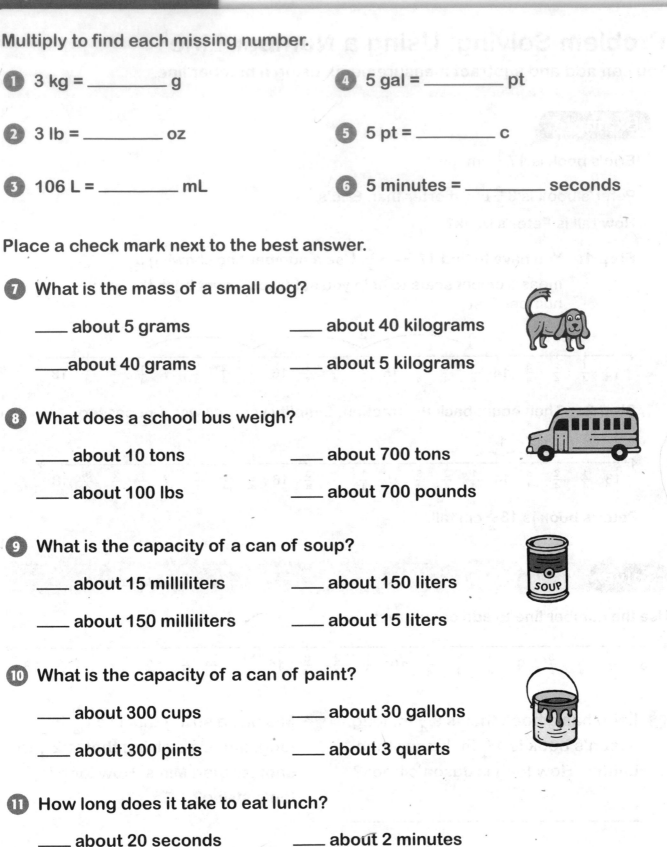

The line plot shows the volume of water in liters in nine small fish tanks.

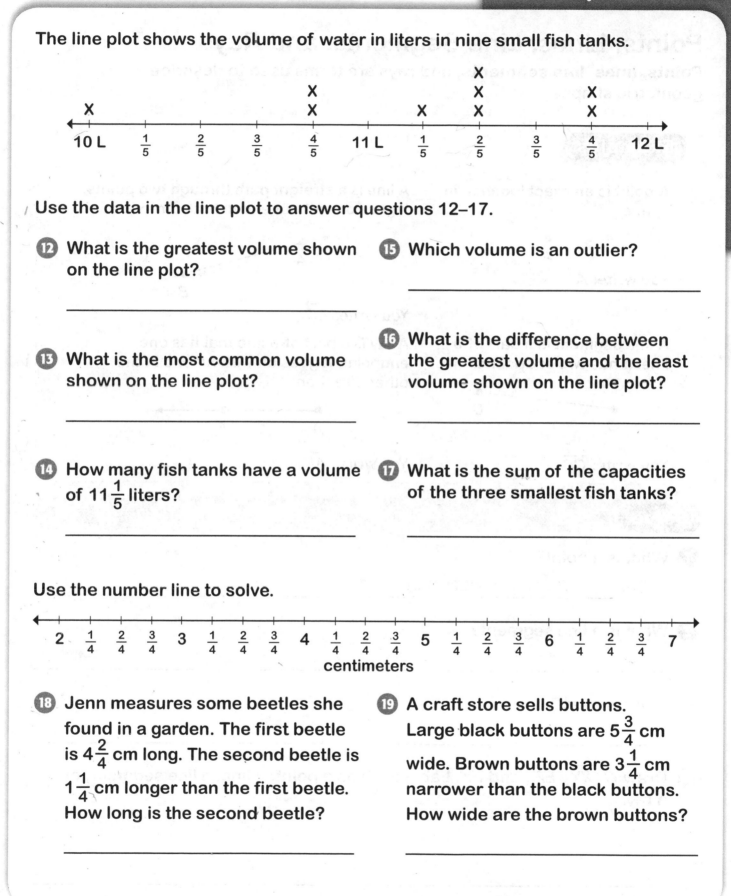

Use the data in the line plot to answer questions 12–17.

12 What is the greatest volume shown on the line plot?

_____

13 What is the most common volume shown on the line plot?

_____

14 How many fish tanks have a volume of $11\frac{1}{5}$ liters?

_____

15 Which volume is an outlier?

_____

16 What is the difference between the greatest volume and the least volume shown on the line plot?

_____

17 What is the sum of the capacities of the three smallest fish tanks?

_____

Use the number line to solve.

centimeters

18 Jenn measures some beetles she found in a garden. The first beetle is $4\frac{2}{4}$ cm long. The second beetle is $1\frac{1}{4}$ cm longer than the first beetle. How long is the second beetle?

_____

19 A craft store sells buttons. Large black buttons are $5\frac{3}{4}$ cm wide. Brown buttons are $3\frac{1}{4}$ cm narrower than the black buttons. How wide are the brown buttons?

_____

Points, Lines, Line Segments, and Rays

Points, lines, line segments, and **rays** are terms used to describe geometric shapes.

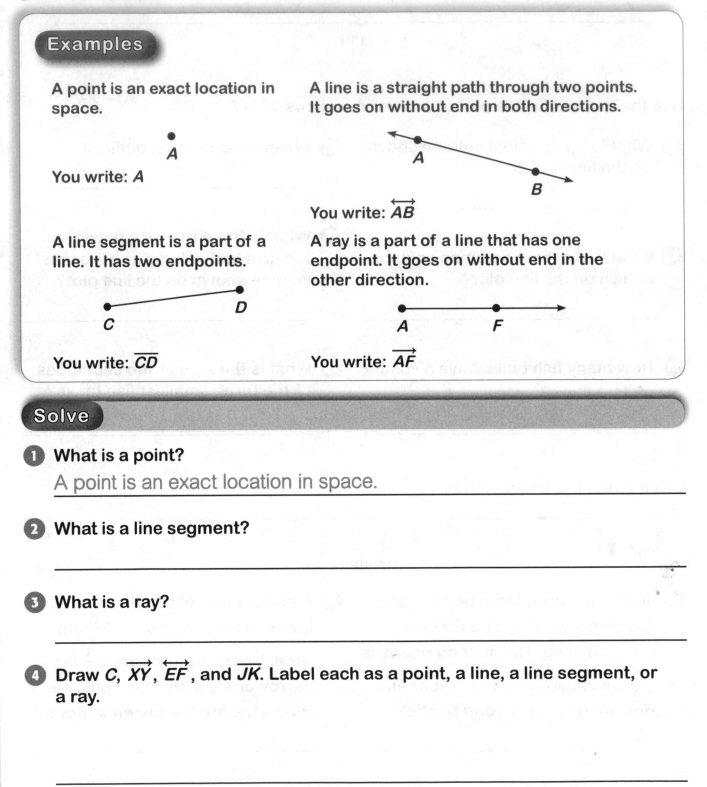

Examples

A point is an exact location in space.

•
A

You write: *A*

A line is a straight path through two points. It goes on without end in both directions.

A

B

You write: $\overleftrightarrow{AB}$

A line segment is a part of a line. It has two endpoints.

C   D

You write: $\overline{CD}$

A ray is a part of a line that has one endpoint. It goes on without end in the other direction.

A   F

You write: $\overrightarrow{AF}$

Solve

1 **What is a point?**

A point is an exact location in space.

2 **What is a line segment?**

_____

3 **What is a ray?**

_____

4 Draw *C*, $\overrightarrow{XY}$, $\overleftrightarrow{EF}$, and $\overline{JK}$. Label each as a point, a line, a line segment, or a ray.

_____

Angles

An angle is a common geometric shape. It is a figure formed by two rays. The rays have the same endpoint.

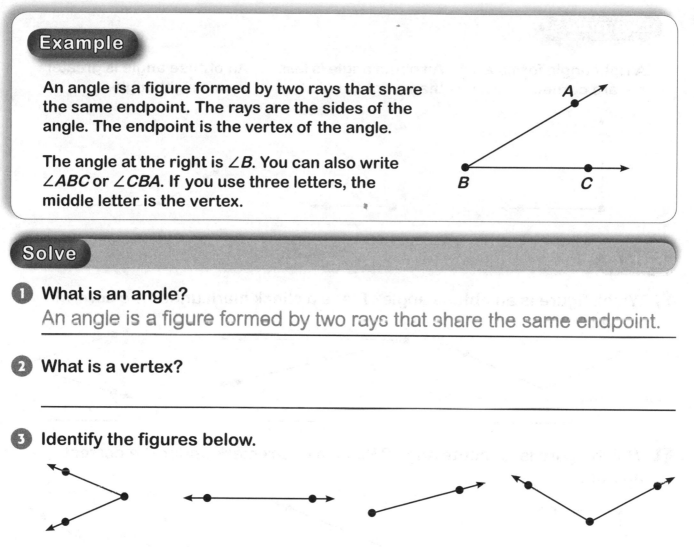

Example

An angle is a figure formed by two rays that share the same endpoint. The rays are the sides of the angle. The endpoint is the vertex of the angle.

The angle at the right is ∠B. You can also write ∠ABC or ∠CBA. If you use three letters, the middle letter is the vertex.

Solve

1 What is an angle?

An angle is a figure formed by two rays that share the same endpoint.

2 What is a vertex?

_____

3 Identify the figures below.

_____   _____   _____   _____

4 Draw ∠EFG. Be sure to write the letters E, F, and G at the correct points.

Name _____

Acute, Right, and Obtuse Angles

Angles have names depending on the size of the opening between the rays.

Examples

A right angle forms a square corner.

An acute angle is less than a right angle.

An obtuse angle is greater than a right angle.

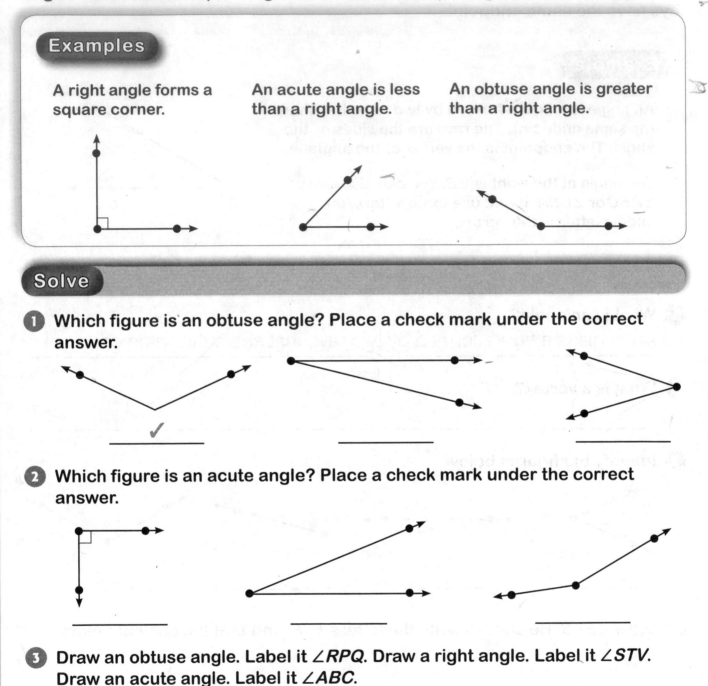

Solve

1. Which figure is an obtuse angle? Place a check mark under the correct answer.

_____ ✓ _____ _____

2. Which figure is an acute angle? Place a check mark under the correct answer.

_____ _____ _____

3. Draw an obtuse angle. Label it ∠RPQ. Draw a right angle. Label it ∠STV. Draw an acute angle. Label it ∠ABC.

Parallel and Perpendicular Lines

Pairs of lines have names depending on if they cross and how they cross.

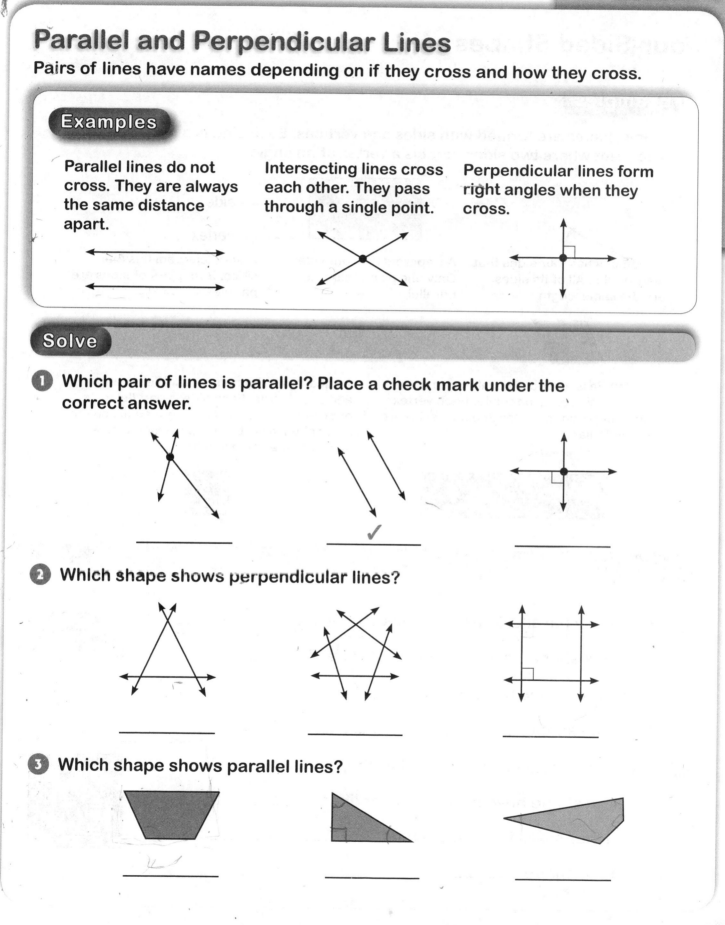

Examples

Parallel lines do not cross. They are always the same distance apart.

Intersecting lines cross each other. They pass through a single point.

Perpendicular lines form right angles when they cross.

Solve

1 Which pair of lines is parallel? Place a check mark under the correct answer.

_____   _____ ✓   _____

2 Which shape shows perpendicular lines?

_____   _____   _____

3 Which shape shows parallel lines?

_____   _____   _____

Four-Sided Shapes _____

Many shapes are formed with sides and vertices. Each side is a line segment. The point where two sides meet is a vertex of an angle.

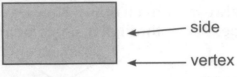

— side

— vertex

A rhombus has four sides that are parallel. All of its sides are the same length.

A trapezoid has four sides. Only one pair of sides is parallel.

A parallelogram has four sides. Both pairs of sides are parallel.

A rectangle is a special parallelogram. Each side is parallel to its opposite. Each vertex is a right angle, so the sides at each vertex are perpendicular.

A square is a rhombus, a parallelogram, and a rectangle. Each side is parallel to its opposite. Each vertex is a right angle, so the sides at each vertex are perpendicular. All sides are the same length.

Solve

❶ How many parallel lines does this shape have? _____2_____

Does this shape have perpendicular lines? _____

Are all four sides the same length? _____

Name the shape. _____

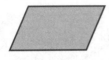

❷ How many parallel lines does this shape have? _____

Does the shape have perpendicular lines? _____

Are all four sides the same length? _____

Name the shape. _____

Triangles

Triangles are classified by the presence or absence of angles of a specific size.

Examples

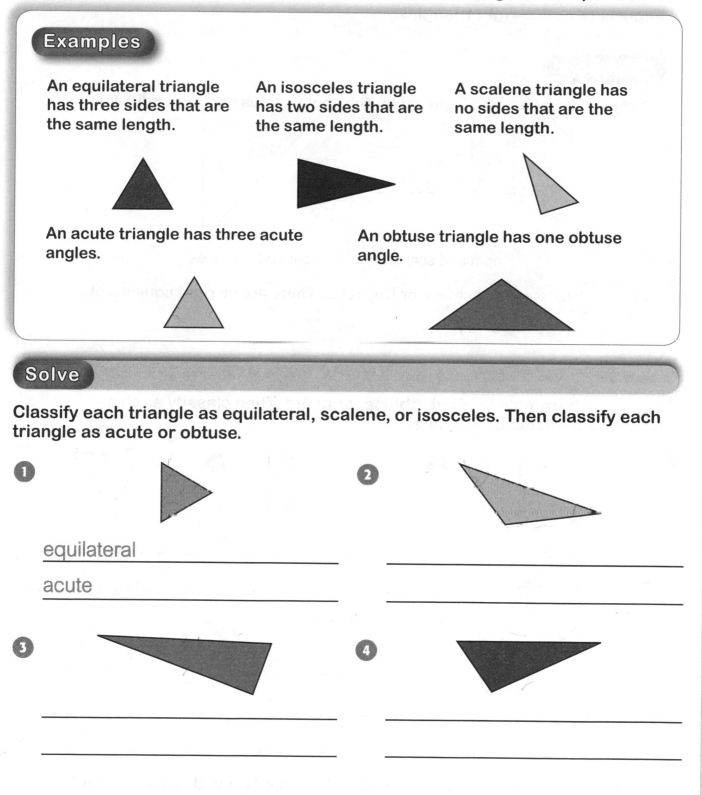

An equilateral triangle has three sides that are the same length.

An isosceles triangle has two sides that are the same length.

A scalene triangle has no sides that are the same length.

An acute triangle has three acute angles.

An obtuse triangle has one obtuse angle.

Solve

Classify each triangle as equilateral, scalene, or isosceles. Then classify each triangle as acute or obtuse.

①

equilateral

acute

②

_____

_____

③

_____

_____

④

_____

_____

Name _____

Right Triangles

Triangles with a right angle have two sides that are perpendicular. These triangles are called right triangles.

Examples

A right triangle has one right angle. The angle's sides are perpendicular.

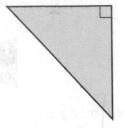

right and scalene   right and isosceles

All right triangles are scalene or isosceles. There are no right equilateral triangles.

Solve

Classify each triangle as right, obtuse, or acute. Then classify each as equilateral, scalene, or isosceles.

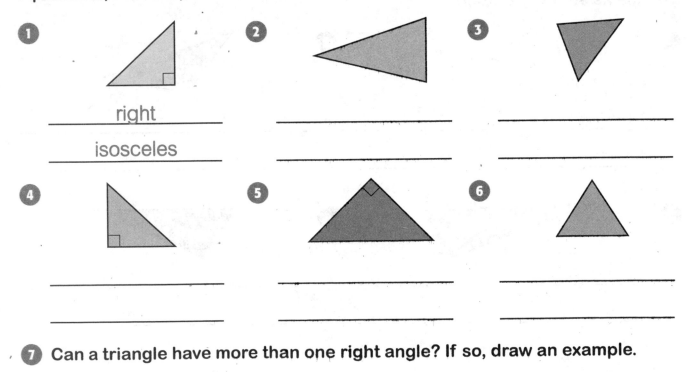

1

_____ right _____

_____ isosceles _____

2

_____

_____

3

_____

_____

4

_____

_____

5

_____

_____

6

_____

_____

7 Can a triangle have more than one right angle? If so, draw an example.

_____

Line Symmetry

A figure has line symmetry if it can be folded in half so the two parts exactly match. The fold line is called a line of symmetry.

Some shapes have one line of symmetry.

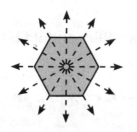

Some shapes have more than one line of symmetry.

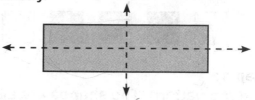

Some shapes have many lines of symmetry.

Some figures have no lines of symmetry.

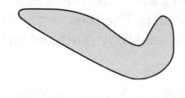

Solve

Look at each shape. Tell how many lines of symmetry each figure has.

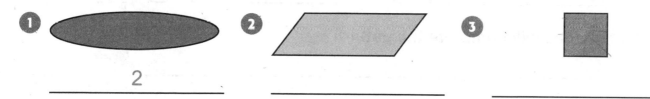

1 _____ 2 _____

2 _____

3 _____

Draw lines of symmetry on each shape if you can.

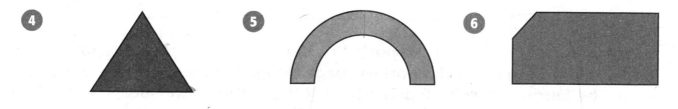

4

5

6

Name _____

Problem Solving: Finding Patterns

Finding patterns can help you solve some problems.

Example

Alex is designing a mural for a wall at his school. He arranges tiles in a pattern. Which color tiles should he use for the unfinished section if he wants to continue the pattern?

Step 1:
Find the pattern. The shapes are all squares. They are all the same size. The colors are different. Look for a color pattern. The tiles are yellow, red, blue, and blue. This pattern repeats.

Step 2:
Continue the pattern:

yellow, red, blue, blue, yellow, red, blue

The last two tiles should be red and blue.

Solve

1 Jenny draws this pattern.

What shapes will complete the pattern?

_____

2 Nasif starts with a number. He adds 920, multiplies by 3, subtracts 1,201, and then divides by 7. The result is 239. What number did Nasif start with?

_____

3 A college teacher is reading reports from her students. The first report is 3 pages long. The second report is 6 pages long. The third report is 9 pages long. The fourth report is 12 pages long. If the pattern continues, how many pages will be in the eighth report?

_____

Name _____

1 What is a ray?

_____

2 What is an angle?

_____

3 Which figure is an obtuse angle? Place a check mark below the correct answer.

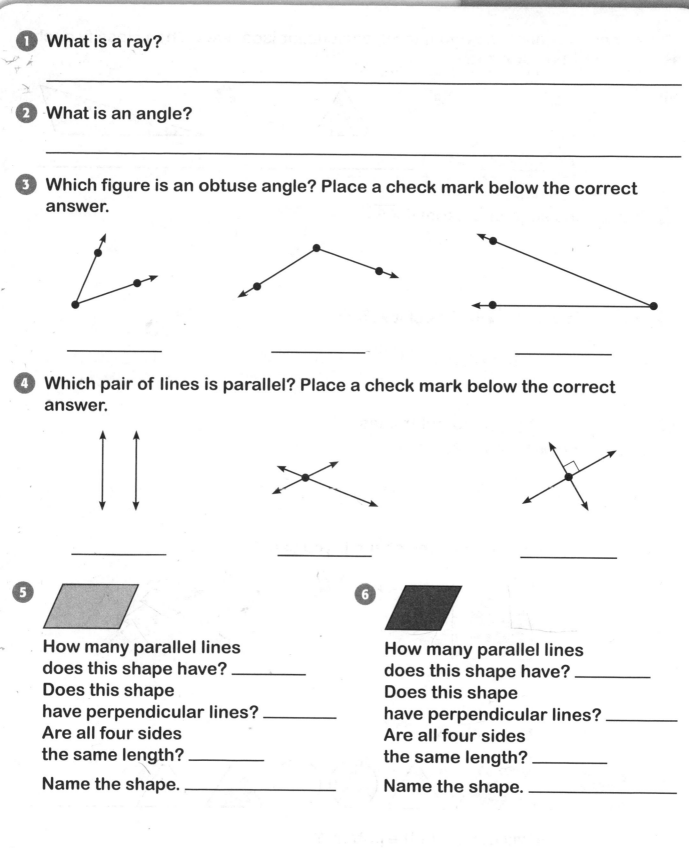

_____   _____   _____

4 Which pair of lines is parallel? Place a check mark below the correct answer.

_____   _____   _____

5 How many parallel lines does this shape have? _____
Does this shape have perpendicular lines? _____
Are all four sides the same length? _____

Name the shape. _____

6 How many parallel lines does this shape have? _____
Does this shape have perpendicular lines? _____
Are all four sides the same length? _____

Name the shape. _____

Classify each triangle as equilateral, scalene, or isosceles. Then classify each as acute, obtuse, or right.

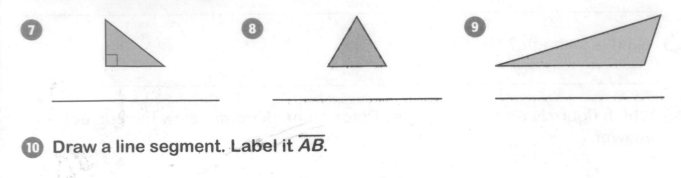

7 _____

8 _____

9 _____

10 Draw a line segment. Label it $\overline{AB}$.

11 Draw an obtuse angle. Label it $\angle RPQ$.

12 Draw a pair of perpendicular lines.
Label them $\overleftrightarrow{MN}$ and $\overleftrightarrow{OP}$.

Draw lines of symmetry on each shape if you can.

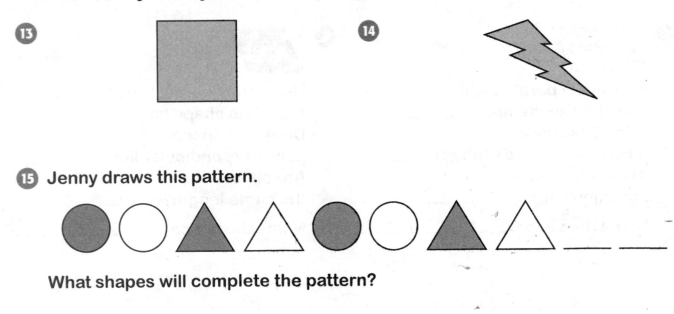

13

14

15 Jenny draws this pattern.

What shapes will complete the pattern?

_____

Angles and Circles

You can measure angles. Angles are measured in units called **degrees**.

Examples

Look at the angle in the circle. The exact center of the circle is the endpoint.

The angle gets larger as you move one ray around the circle. The curved gap between the rays can be measured.

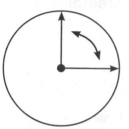

An angle that turns $\frac{1}{360}$ of a circle has an angle measure of one degree.

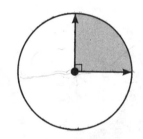

An angle that turns through $\frac{90}{360}$ of a circle is a right angle. It has an angle measure of 90 degrees.

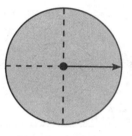

An angle that turns through $\frac{360}{360}$ of a circle has made a full rotation. It is made of four 90-degree angles and has an angle measure of 360 degrees.

Solve

Write each missing number.

1 1 right angle = _____90_____

3 3 right angles = _____

2 2 right angles = _____

4 4 right angles = _____

5 A 90-degree angle is a right angle.
What would a 45-degree angle look like?
Draw it.

Name _____

Measuring Angles

Examples

A protractor is a math tool used to measure angles. How do you use a protractor?

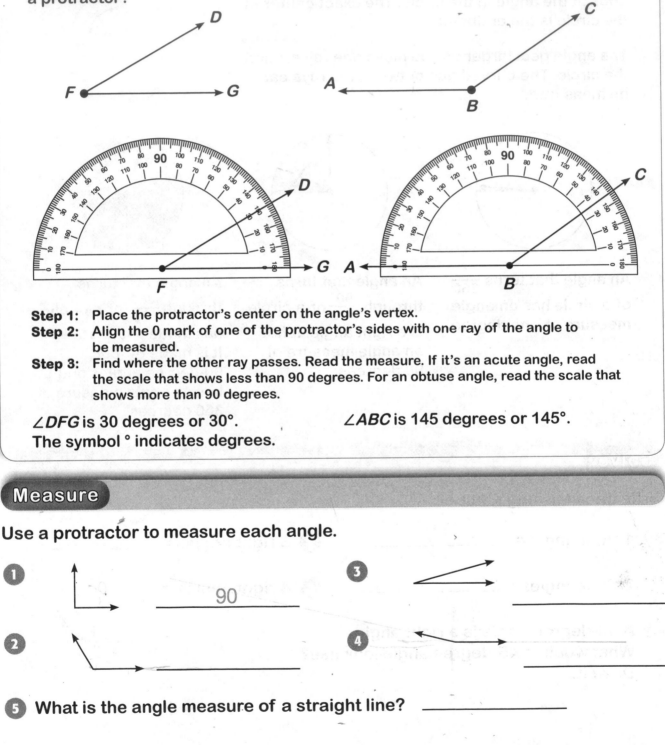

Step 1: Place the protractor's center on the angle's vertex.
Step 2: Align the 0 mark of one of the protractor's sides with one ray of the angle to be measured.
Step 3: Find where the other ray passes. Read the measure. If it's an acute angle, read the scale that shows less than 90 degrees. For an obtuse angle, read the scale that shows more than 90 degrees.

∠DFG is 30 degrees or 30°.
The symbol ° indicates degrees.

∠ABC is 145 degrees or 145°.

Measure

Use a protractor to measure each angle.

1 _____90_____

3 _____

2 _____

4 _____

5 What is the angle measure of a straight line? _____

Drawing Angles

You can use a protractor to draw angles too.

Example

How do you draw a 100° angle?

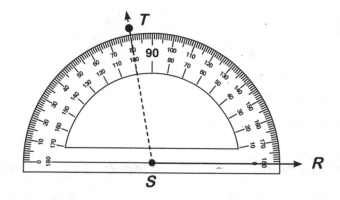

Step 1:   Draw a ray. Label it $\overrightarrow{SR}$.
Step 2:   Place the center mark of the protractor on point *S*. Align the 0° mark with the ray.
Step 3:   Mark a point at 100°. Label the point *T*.
Step 4:   Draw a ray from the vertex through the point you drew. Use the straight edge of the protractor or a ruler to help you draw a straight line.

∠*TSR* is a 100° angle.

Draw

Draw an angle with each measure. Use a protractor.

1 130°

3 65°

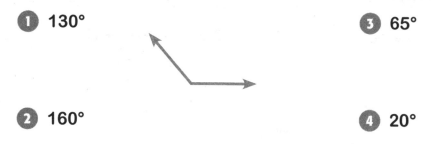

2 160°

4 20°

5 Look at each angle you drew. Classify each as acute, obtuse, or right.

#1 _____

#3 _____

#2 _____

#4 _____

Name _____

Adding Angle Measures

You can add angle measures to solve problems.

Example

∠*NLM* measures 94°. ∠*MLK* measures 32°. What is the measure of ∠*NLK*?

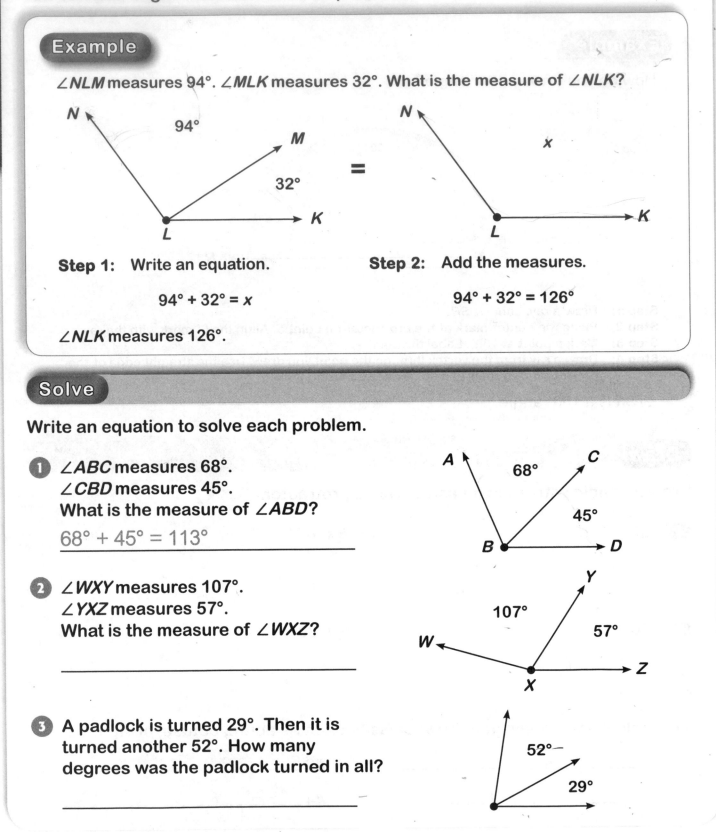

Step 1: Write an equation.

$$94° + 32° = x$$

Step 2: Add the measures.

$$94° + 32° = 126°$$

∠*NLK* measures 126°.

Solve

Write an equation to solve each problem.

1 ∠*ABC* measures 68°.
∠*CBD* measures 45°.
What is the measure of ∠*ABD*?

$\underline{68° + 45° = 113°}$

2 ∠*WXY* measures 107°.
∠*YXZ* measures 57°.
What is the measure of ∠*WXZ*?

_____

3 A padlock is turned 29°. Then it is
turned another 52°. How many
degrees was the padlock turned in all?

_____

Subtracting Angle Measures

You can subtract angle measures to solve problems too.

Example

∠QRT measures 152°. If ∠SRT measures 64°, what is the measure of ∠QRS?

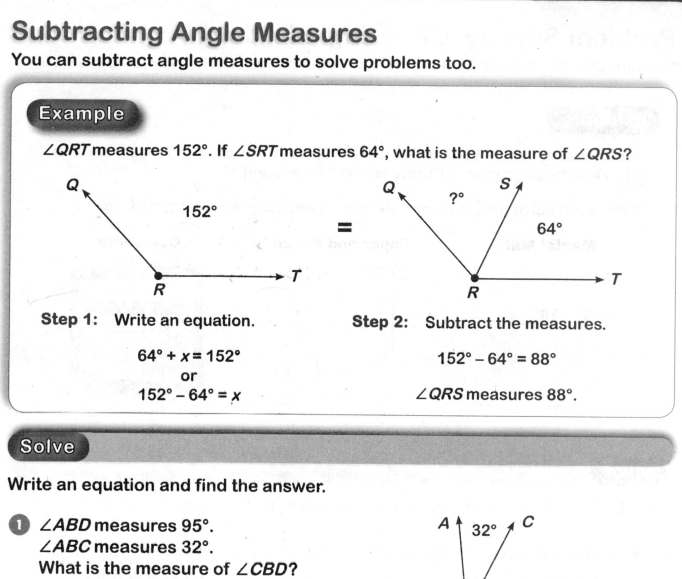

Step 1: Write an equation.

$$64° + x = 152°$$
or
$$152° - 64° = x$$

Step 2: Subtract the measures.

$$152° - 64° = 88°$$

∠QRS measures 88°.

Solve

Write an equation and find the answer.

1 ∠ABD measures 95°.
∠ABC measures 32°.
What is the measure of ∠CBD?

$95° - 32° = 63°$ or $32° + x = 95°$
_____

2 ∠EFH measures 151°.
∠EFG measures 77°.
What is the measure of ∠GFH?

_____

3 A lock is set at 0. Its arrow is turned 117 degrees forward and 23 degrees backward Then it is turned forward 54 degrees. At what degree measure is the arrow pointing?

_____

Name _____

Problem Solving: Choosing a Method

You can use mental math, paper and pencil, or calculators to solve problems.

Example

Mike draws an angle that measures 149°. Gina draws an angle that measures 78°. How much greater is the measure of Mike's angle?

Before you solve the problem, you need to decide which method to use.

Mental Math	Paper and Pencil	Calculator

Solve

Tell which method you chose for each problem. Explain why.

1 Kameko and Bill have the same size water bottle. Kameko's bottle is $\frac{3}{10}$ full of water. Bill's bottle is $\frac{5}{10}$ full of water. How much more water does Bill's bottle have?

$\frac{2}{10}$ more or $\frac{1}{5}$ more; I used mental math. It's easy to subtract $\frac{3}{10}$

from $\frac{5}{10}$ in your head.

2 A mechanic turns a bolt 89°. Then he turns it another 67°. How many degrees did he turn the bolt altogether?

_____

3 A small office building has a rectangular parking lot. It measures 58 m long and 52 m wide. What is the perimeter of the parking lot?

_____

Write each missing number.

1 2 right angles = _____

2 3 right angles = _____

Use a protractor to measure each angle.

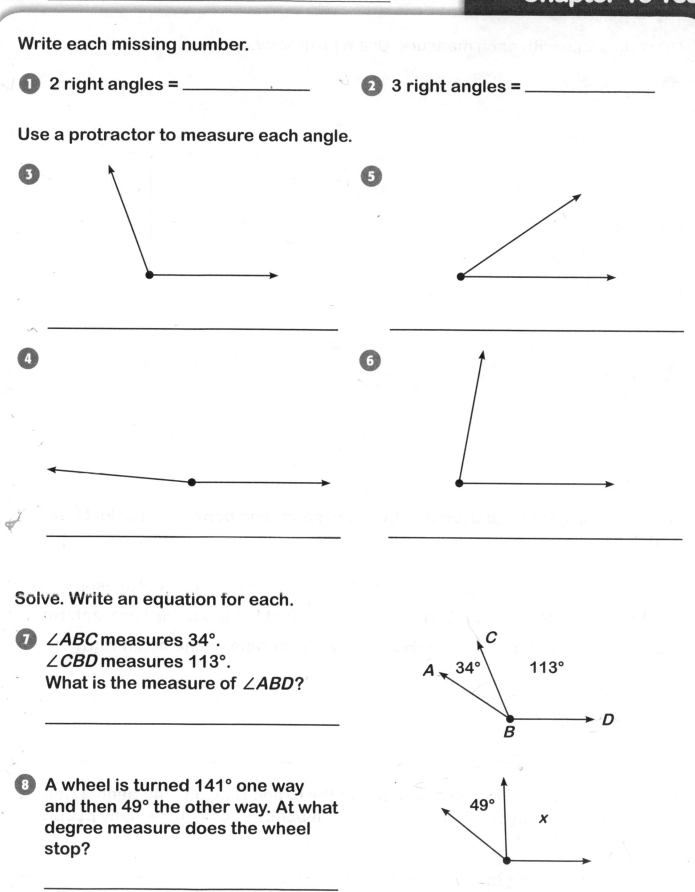

3

_____

5

_____

4

_____

6

_____

Solve. Write an equation for each.

7 ∠ABC measures 34°.
∠CBD measures 113°.
What is the measure of ∠ABD?

_____

8 A wheel is turned 141° one way
and then 49° the other way. At what
degree measure does the wheel
stop?

_____

Draw an angle with each measure. Use a protractor.

9 50°

10 145°

11 97°

Solve. Tell whether you used mental math, paper and pencil, or a calculator. Explain why.

12 Connor has a roll of tape. He uses $2\frac{1}{7}$ ft to tape one present. He uses $4\frac{2}{7}$ ft to tape another present. Then he uses $3\frac{2}{7}$ feet to tape a third present. He has $14\frac{1}{7}$ ft of tape left over. How many feet of tape did he begin with?

_____

_____

13 A school group goes on a field trip to the Grand Canyon. 140 people ride 7 buses. If each bus holds the same number of people, how many people are on each bus?

_____

Fill in each blank. Use the best term from the word bank. You may not use every term.

Word Bank

• even	• multiples	• odd
• rhombus	• factors	• number line
• quotient	• sum	• divisor
• difference	• skip count	• product
• add	• subtract	• array
• equal groups	• multiply	• divide
• circle	• square	• triangle
• estimate	• fraction	• expression
• denominator	• numerator	• equation

1 A(n) _____ is a group of numbers with an operation symbol.

2 To find the difference between two numbers, you _____.

3 You _____ to find out how many in all.

4 The _____ is the number below the fraction bar in a fraction.

5 A(n) _____ has four equal sides and four equal angles.

6 A(n) _____ is the number of equal groups a dividend is divided by in a division problem.

7 An array is a good way to show _____ of objects.

8 A(n) _____ is a shape that has zero sides and zero angles.

9 You _____ to find a number close to an exact amount.

10 You _____ to find how many equal groups a number can be divided into.

11 The answer to an addition problem is called the _____.

12 You can _____ two numbers together to find the product.

13 A(n) _____ is a symbol used to name part of a whole.

14 The _____ is the answer to a division problem.

Name _____

Write each number in standard form.

15 7 tens

_____

16 50 + 1

_____

17 seventy-seven

_____

18 nineteen

_____

19 7 hundreds 1 ten 4 ones

_____

20 800 + 8

_____

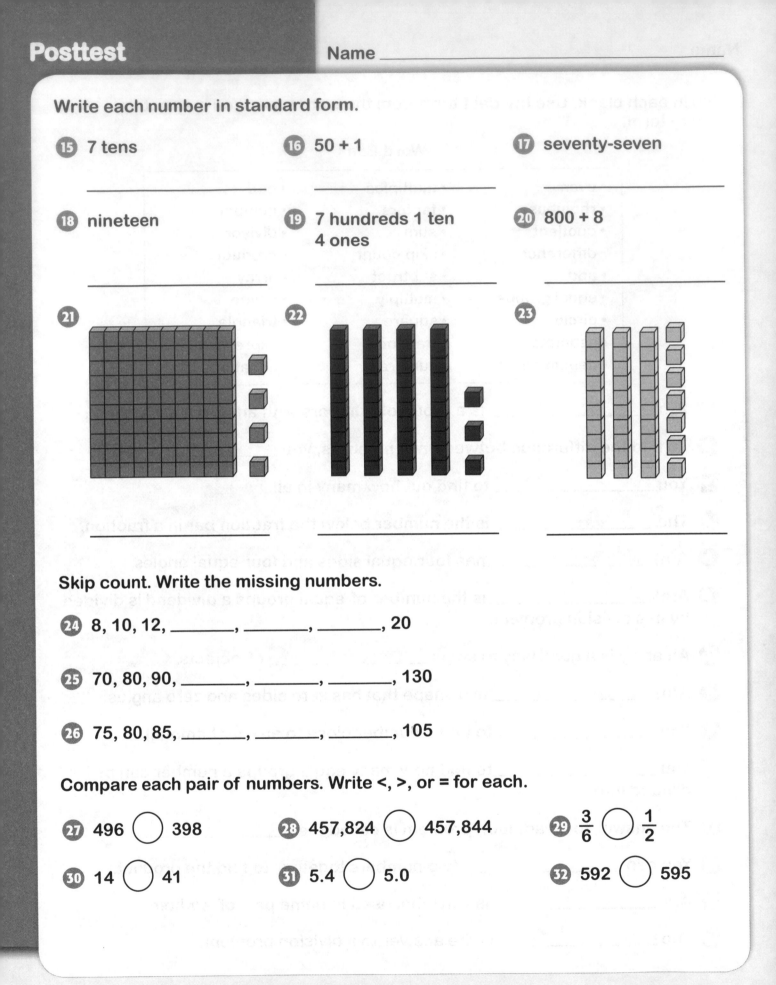

21

_____

22

_____

23

_____

Skip count. Write the missing numbers.

24 8, 10, 12, _____, _____, _____, 20

25 70, 80, 90, _____, _____, _____, 130

26 75, 80, 85, _____, _____, _____, 105

Compare each pair of numbers. Write <, >, or = for each.

27 496 ◯ 398

28 457,824 ◯ 457,844

29 $\frac{3}{6}$ ◯ $\frac{1}{2}$

30 14 ◯ 41

31 5.4 ◯ 5.0

32 592 ◯ 595

Name _____

Add. Write the sum.

33
```
   356
+  871
```

34
```
  18437
+ 31854
```

35
```
  36792
+  9654
```

36
```
  63121
+ 42190
```

37
```
   562
   711
+  631
```

38
```
  25582
  72990
+ 31854
```

39
```
   5612
   6453
+  8904
```

40
```
   953
   110
+  872
```

Subtract. Write the difference.

41
```
  529228
- 337643
```

42
```
  815027
-  56774
```

43
```
  437184
-  72138
```

44
```
  90543
- 90251
```

Round to the nearest thousand to estimate the sum or difference.

45 82,019 – 78,901 = _____   **46** 351,092 + 238,091 = _____

Round to the nearest hundred to estimate the sum or difference.

47 74,521 + 676,890 = _____   **48** 90,932 – 65,721 = _____

Solve.

49 A rectangular yard is 8 meters long and 5 meters wide. What is the perimeter of the yard?

_____

50 A sandbox measures 7 feet long and 5 feet wide. What is the area of the sandbox?

_____

Find the product.

51 4 × 0 = _____ **52** 2 × 8 = _____ **53** 9 × 8 = _____

54 12 × 4 = _____ **55** 10 × 5 = _____ **56** 5 × 1 = _____

Find the quotient.

57 12 ÷ 3 = _____ **58** 32 ÷ 4 = _____ **59** 35 ÷ 7 = _____

60 132 ÷ 12 = _____ **61** 88 ÷ 11 = _____ **62** 60 ÷ 10 = _____

63 Which multiplication property is shown in this sentence?
5 × (12 + 15) = (5 × 12) + (5 × 15)

_____

Identify each number as prime or composite.

64 6 _____ **66** 31 _____

65 11 _____ **67** 66 _____

Find all the factor pairs of each number.

68 12 _____

69 19 _____

Solve. Write the expressions you used and the answers.

70 Ms. Neal is making a large pot of stew for a family reunion. The recipe calls for 4 times as many peas as carrots. If she uses 2 cups of carrots, how many cups of peas will she need?

_____

71 A deli has 6 times as many bottles of juice as pickles. If the deli has 54 bottles of juice, how many pickles are in the deli? Use the variable *y* for the number of pickles.

_____

Write an equivalent fraction for each.

72 $\frac{3}{4}$ _____   73 $\frac{24}{40}$ _____   74 $\frac{4}{36}$ _____

Add or subtract. Write your answers in simplest form.

75 $\frac{2}{5} + \frac{1}{5} =$ _____   76 $\frac{3}{10} + \frac{5}{10} =$ _____   77 $78\frac{9}{15} - 47\frac{5}{15} =$ _____

Write each fraction as a decimal.

78 $\frac{5}{10}$ _____   79 $\frac{30}{100}$ _____   80 $\frac{78}{100}$ _____   81 $\frac{40}{50}$ _____

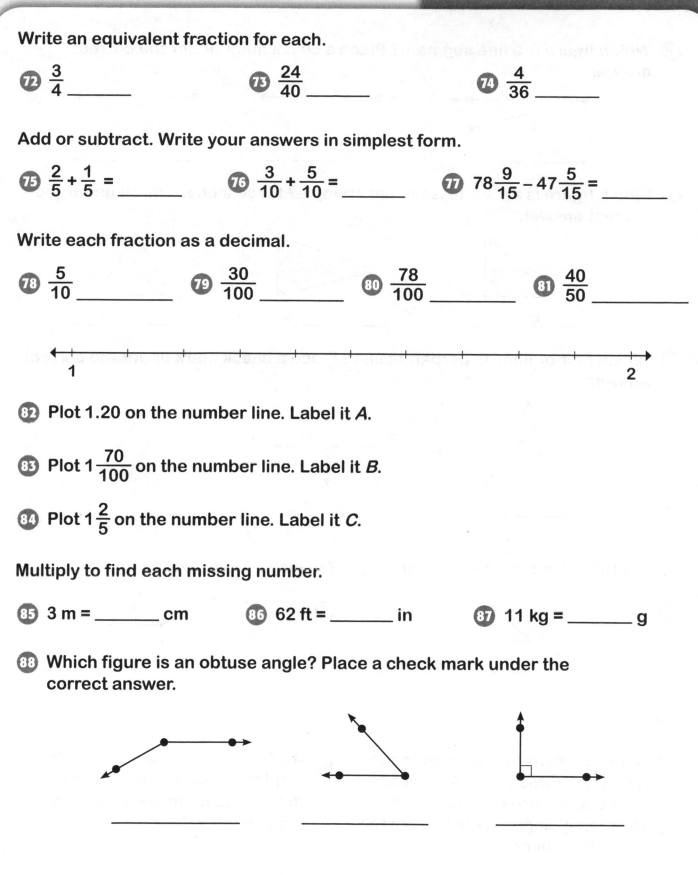

82 Plot 1.20 on the number line. Label it *A*.

83 Plot $1\frac{70}{100}$ on the number line. Label it *B*.

84 Plot $1\frac{2}{5}$ on the number line. Label it *C*.

Multiply to find each missing number.

85 3 m = _____ cm   86 62 ft = _____ in   87 11 kg = _____ g

88 Which figure is an obtuse angle? Place a check mark under the correct answer.

_____   _____   _____

89 Which figure is a line segment? Place a check mark under the correct answer.

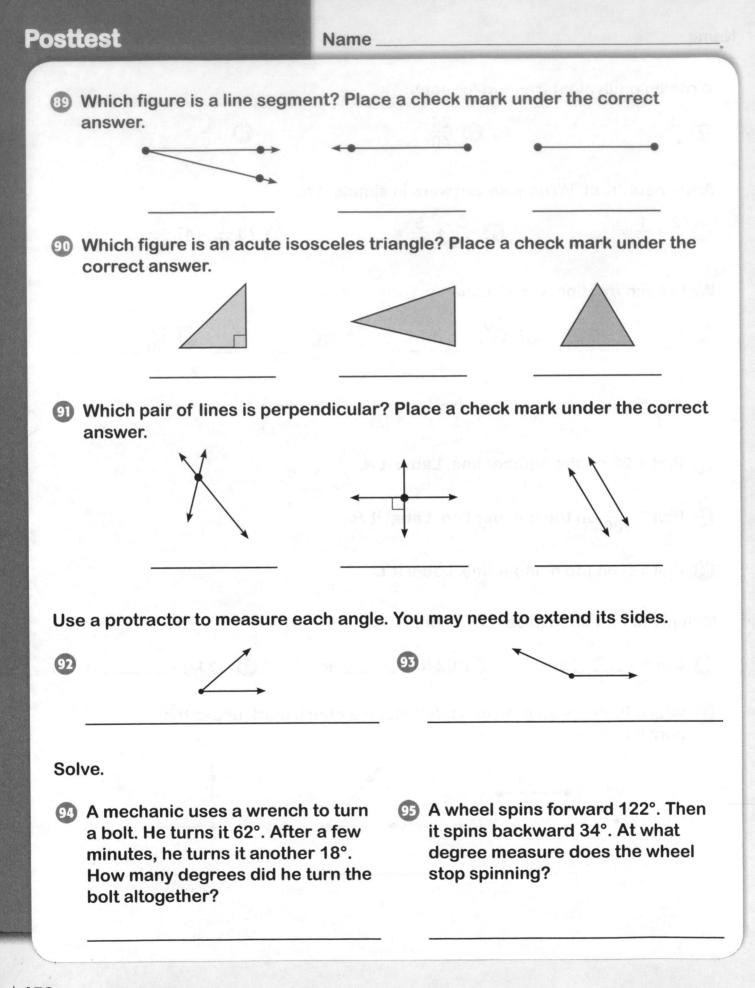

_____   _____   _____

90 Which figure is an acute isosceles triangle? Place a check mark under the correct answer.

_____   _____   _____

91 Which pair of lines is perpendicular? Place a check mark under the correct answer.

_____   _____   _____

Use a protractor to measure each angle. You may need to extend its sides.

92

93

_____   _____

Solve.

94 A mechanic uses a wrench to turn a bolt. He turns it 62°. After a few minutes, he turns it another 18°. How many degrees did he turn the bolt altogether?

95 A wheel spins forward 122°. Then it spins backward 34°. At what degree measure does the wheel stop spinning?

_____   _____

Acute Angle: An angle that is less than 90 degrees. *(p. 128)*

Acute Triangle: A triangle with three acute angles. *(p. 131)*

Addend: The numbers you add to find a sum. *(p. 22)*

Additive Comparison: A comparison using addition expressions. *(p. 28)*

Angle: A figure formed by two rays that share the same endpoint. *(p. 127)*

Area: The number of square units needed to cover a region. *(p. 109)*

Array: A way to show equal groups of objects in columns and rows. *(p. 32)*

Associative Property of Addition: You can change the grouping of addends. The sum remains the same. *(p. 22)*

Associative Property of Multiplication: You can change the grouping of factors. The product remains the same. *(p. 37)*

Capacity: The amount a container can hold measured in liquid units. *(p. 118)*

Centimeter: A metric unit used to measure length. *(p. 105)*

Commutative Property of Addition: You can add numbers in any order. The sum remains the same. *(p. 22)*

Commutative Property of Multiplication: You can multiply factors in any order. The product remains the same. *(p. 36)*

Compatible Numbers: Numbers that are easy to use mathematically. *(p. 59)*

Composite Number: A whole number greater than 1 that has more than two factors. *(p. 45)*

Cup: A customary unit of capacity. *(p. 119)*

Decimal Number: A number with one or more digits to the right of a decimal point. *(p. 96)*

Decimal Point: A dot used to separate ones from tenths in a decimal number. *(p. 96)*

Degree (°): A unit of measurement for angles. *(p. 137)*

Denominator: The total number of equal parts of a whole or a set. The number below the fraction bar in a fraction. *(p. 72)*

Digit: The symbols 0, 1, 2, 3, 4, 5, 6, 7, 8, and 9 that are used to write numbers. *(p. 14)*

Distributive Property of Multiplication: You can multiply two addends by a factor. You can also multiply each addend by the same factor and add the products. The final total is the same. *(p. 37)*

Dividend: The number divided in a division problem. *(p. 38)*

Divisor: The number of equal groups a dividend is divided by in a division problem. *(p. 38)*

Endpoint: A point at the beginning of a ray or at either side of a line segment. *(p. 126)*

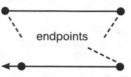

endpoints

Equilateral Triangle: A triangle in which all sides are the same length. *(p. 131)*

Equivalent Fractions: Fractions that name the same part of a whole. *(p. 73)*

$$\frac{1}{2} = \frac{3}{6}$$

Estimate: A number close to an exact amount. A careful guess. *(p. 29)*

Expanded Form: A number form that shows the value of the digits in a number. *(p. 15)*

$$437 = 400 + 30 + 7$$

Expression: A number or a group of numbers with an operation symbol. Some expressions have variables. *(p. 27)*

Factors: The numbers you multiply in a multiplication problem. *(p. 32)*

Foot, Feet: A customary unit used to measure length. *(p. 107)*

Fraction: Symbols, such as $\frac{1}{2}$ or $\frac{2}{3}$, used to name part of a whole or parts of a set. *(p. 81)*

Gallon: A customary unit used to measure capacity. *(p. 119)*

Gram: A metric unit used to measure mass. *(p. 116)*

Identity Property of Addition: Adding zero to a number does not change the number. *(p. 22)*

Identity Property of Multiplication: Any number multiplied by 1 equals the number. *(pp. 34, 36)*

Improper Fraction: A fraction with a numerator greater than or equal to the denominator. *(p. 84)*

Inch: A customary unit used to measure length. *(p. 107)*

Intersecting Lines: Lines that cross at a single point. *(p. 129)*

Inverse Operations: Operations that undo each other. *(pp. 23, 39)*

Isosceles Triangle: A triangle that has two sides of equal length. *(p. 131)*

Kilogram: A metric unit used to measure mass. *(p. 116)*

Kilometer: A metric unit used to measure length. *(p. 105)*

Line: A straight path through two points. It goes on without end in both directions. *(p. 126)*

Line of Symmetry: A line on which a figure can be folded so both halves exactly match. *(p. 133)*

Line Plot: A way to show data with a number line. *(p. 121)*

Line Segment: Part of a line with two endpoints. *(p. 126)*

Liter: A metric unit used to measure capacity. *(p. 118)*

Mass: The measure of the amount of matter in an object. *(p. 116)*

Meter: A metric unit used to measure length. *(p. 105)*

Mile: A customary unit used to measure length. *(p. 107)*

Milliliter: A metric unit used to measure capacity. *(p. 118)*

Mixed Number: A number containing a whole number and a fraction. *(p. 84)*

Multiple: The product of two whole numbers. *(p. 47)*

Multiplication Comparison: A comparison using multiplication expressions. *(p. 48)*

Numerator: A specific number of equal parts of a whole or a set. The number above the fraction bar in a fraction. *(p. 72)*

Obtuse Angle: An angle that is more than 90 degrees. *(p. 128)*

Glossary

Obtuse Triangle: A triangle with one obtuse angle. *(p. 131)*

Ounce: A customary unit of weight. *(p. 117)*

Outlier: Any number that is very different from the rest of the numbers in a data set. *(p. 121)*

Parallel Lines: Lines that are always the same distance apart and never cross. *(p. 129)*

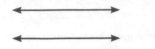

Parallelogram: A four-sided figure in which opposite sides are parallel. *(p. 130)*

Partial Products: Products found by breaking a factor into ones, tens, hundreds, and so on and then multiplying each of these by another factor. *(p. 54)*

Perimeter: The number of units around a figure. *(p. 110)*

Perpendicular Lines: Lines that form right angles when they cross. *(p. 129)*

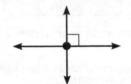

Pint: A customary unit used to measure capacity. *(p. 119)*

Place Value: The value of a digit. *Example:* The value of 4 in 5,429 is 400. *(p. 14)*

Point: An exact location in space. *(p. 126)*

A

Pound: A customary unit used to measure weight. *(p. 117)*

Prime Number: A whole number greater than 1 that has only two factors, 1 and itself. *(p. 45)*

Product: The answer to a multiplication problem. *(p. 32)*

Protractor: A math tool used to measure angles. *(p. 138)*

Quart: A customary unit used to measure capacity. *(p. 119)*

Quotient: The answer to a division problem. *(p. 38)*

Ray: A part of a line with one endpoint. It goes on without end in the other direction. *(p. 126)*

Regroup: To exchange equal amounts using place value. *(p. 24)*

10 ones = 1 ten

Remainder: The amount left over after division is complete. *(p. 64)*

Rhombus: A four-sided figure in which opposite sides are parallel and all four sides are equal length. *(p. 130)*

Right Angle: An angle that measures exactly 90 degrees. *(p. 128)*

Right Triangle: A triangle with one right angle. *(p. 132)*

Round, Rounding: To express a number to the nearest ten, hundred, thousand, and so on. *(p. 18)*

Scalene Triangle: A triangle with three sides of different lengths. *(p. 131)*

Side (angle): One of the rays in an angle. *(p. 127)*

Side (shape): One of the line segments that makes up a shape. *(p. 130)*

Simplest Form: A fraction in which 1 is the only number that divides both the numerator and the denominator with no remainder. *(p. 81)*

Standard Form: A way to write a number using only digits. *(p. 15)*

2,350

Symmetry, Symmetric: A figure that can be folded in half so the two parts match exactly. *(p. 133)*

Ton: A customary unit used to measure weight. *(p. 117)*

Trapezoid: A four-sided figure in which only one pair of sides is parallel. *(p. 130)*

Variable: A symbol that stands for a number. *(p. 27)*

Vertex (Vertices): The endpoint that two rays share to form an angle. *(pp. 127, 130)*

Weight: The measure of how heavy something is. *(p. 116)*

Word Form: A way to write a number using words. *(p. 15)*

four hundred seventy-eight

Yard: A customary unit used to measure length. *(p. 107)*

Zero Property of Multiplication: Any number multiplied by zero is zero. *(p. 34)*

Pretest
pages 8–13
1. number line
2. odd
3. product
4. round
5. skip count
6. Factors
7. multiples
8. even
9. array
10. quotient
11. difference
12. triangle
13. equation
14. numerator
15. 28   16. 32
17. 416   18. 14
19. 350   20. 57
21. 17   22. 30
23. 24
24. 8, 10, 12
25. 40, 50, 60
26. 40, 45, 50
27. <   28. <
29. =   30. >
31. >   32. <
33. 1,670
34. 143,937
35. 23,572
36. 139,422
37. 2,481
38. 205,970
39. 24,222
40. 1,876
41. 193,881
42. 233,160
43. 850,505
44. 215
45. 15,000
46. 153,000
47. 789,500
48. 82,200
49. 52 m   50. 99 sq ft
51. 0   52. 48
53. 63   54. 39
55. 70   56. 23
57. 41R1   58. 5
59. 7   60. 12
61. 11   62. 11
63. Distributive Property of Multiplication
64. prime
65. composite
66. composite
67. composite
68. 1 and 18, 2 and 9, 3 and 6
69. 1 and 23
70. 6 × 7 = 42 cups of water

71. 12 × y = 96; 96 ÷ 12 = 8 red marbles
72. Possible answer: $\frac{6}{10}$
73. Possible answer: $\frac{5}{11}$
74. Possible answer: $\frac{1}{17}$
75. 1
76. $\frac{3}{4}$
77. 464$\frac{7}{13}$
78. ●
79. ■
80. ■
81. square
82. triangle
83. circle
84. rectangle
85. 55 inches
86. 300 pounds
87. Yes; Possible answer: The Commutative Property of Addition says that the sum remains the same if you change the order of the addends. So 67 + 29 is equal to 29 + 67.
88. No; Possible answer: When you multiply 4 and 5, there is a zero in the product. So when you multiply 4 and 500, you have to add another zero to the ones already there. The answer is 2,000.
89. drums
90. piano
91. 9 students
92. 6 more students

Chapter 1

Chapter 1
Lesson 1, page 14
1. hundreds
2. ones
3. hundred thousands
4. tens
5. ten thousands

6. thousands
7. hundreds
8. ten thousands
9. Answers will vary, but all should be 6-digit numbers with a 4 in the thousands place and a 9 in the tens place. Sample answer: 124,095

Chapter 1
Lesson 2, page 15
1. 7,000 + 400 + 90 + 8
2. 500 + 70 + 1
3. 349,572
4. 72,963
5. five hundred thirty-one thousand, seven hundred ninety-four
6. 99,999; 90,000 + 9,000 + 900 + 90 + 9

Chapter 1
Lesson 3, page 16
1. 500
2. 70
3. 6,000
4. 4
5. 40,000; Possible answer: The place a digit has in a number shows the value of the digit. If I move the digit three places to the left, I can multiply by 10 three times to find the value.

Chapter 1
Lesson 4, page 17
1. <   2. >
3. >   4. <
5. <   6. >
7. =   8. >
9. Possible answers: 640,100; 641,000; 642,000; 640,010

Chapter 1
Lesson 5, page 18
1. 830   2. 780
3. 6,700   4. 73,500
5. 50,000   6. 700,000
7. 250,000   8. 490,530
9. 3,542

Chapter 1
Lesson 6, page 19
1. Ivan's marker is blue. Ester's marker is black. Lloyd's marker is yellow. Cindy's marker is red.

Chapter 1
Test, pages 20–21
1. thousands
2. tens
3. hundred thousands

4. ten thousands
5. hundreds
6. ones
7. 32,761
8. 30,000 + 2,000 + 700 + 60 + 1
9. 800,000 + 20,000 + 3,000 + 500 + 40 + 4
10. eight hundred twenty-three thousand, five hundred forty-four
11. 600,000   12. 300
13. <   14. >
15. <   16. >
17. <   18. =
19. >   20. >
21. 700   22. 62,400
23. 600,000   24. 3,600
25. Bela has 25 points. Jason has 20 points. Tom has 30 points.
26. Julia, Ben, Ali, Lin

Chapter 2

Chapter 2
Lesson 1, page 22
1. 17; Associative Property of Addition
2. 0; Identity Property of Addition
3. 45; Commutative Property of Addition
4. 6; Commutative Property of Addition
5. 24; Associative Property of Addition
6. 23; Identity Property of Addition
7. No. The Identity Property of Addition states that when zero is added to any number, the number does not change.

Chapter 2
Lesson 2, page 23
1. 81; 81 − 39 = 42
2. 25; 25 − 10 = 15
3. 31; 31 − 23 = 8
4. 36; 36 − 4 = 32
5. 26; 26 + 12 = 38
6. 25; 25 + 4 = 29
7. 31; 31 + 9 = 40
8. 12; 12 + 11 = 23
9. 85 − 66 = 19 pages; 19 + 66 = 85

Chapter 2
Lesson 3, page 24
1. 1,055
2. 12,418

Answers

3. 82,259
4. 75,542
5. 585,844
6. 972,474
7. 64,108
8. 697,328
9. 102,664
10. 857,884
11. Possible answer: No. The sum is less than 100,000 so I do not need to regroup numbers to a higher place value.

Chapter 2
Lesson 4, page 25
1. 2,237   2. 20,601
3. 484,781   4. 425,608
5. 180,207   6. 901,247
7. 467,081

Chapter 2
Lesson 5, page 26
1. 434
2. 4,628
3. 33,728
4. 401,365
5. 858,258
6. 26,603
7. 238,231
8. 26,283
9. 44,279
10. 686,089
11. Possible answer: It was not written because the digit 0 has no value. There are no hundred thousands in the answer.

Chapter 2
Lesson 6, page 27
1. 5; 6; 7; 8
2. 12; 24; 36; 48; 60; 72
3. 27; 24; 21; 18; 15; 12
4. It means that 12 is added to each number in the x row.

Chapter 2
Lesson 7, page 28
1. 57 + y = 85, y = 28, 102 + 28 = 130 inches tall
2. 17 + y = 33, y = 16, 28 + 16 = 44 years old

Chapter 2
Lesson 8, page 29
1. about 14,000 people or about 13,400 people
2. about 20 years younger
3. about 250 people

Chapter 2
Test, pages 30–31
1. Identity Property of Addition
2. Commutative Property of Addition
3. Identity Property of Addition
4. Associative Property of Addition
5. Associative Property of Addition
6. Commutative Property of Addition
7. No. The Identity Property of Addition states that when zero is added to any number, the number does not change.
8. 228 – 83
9. 332 + 266
10. 1,187
11. 92,003
12. 60,138
13. 487,812
14. 852,738
15. 17,128
16. 590
17. 2,192
18. 22,384
19. 313,269
20. 689,052
21. 32,488
22. 30; 35; 40; 45; 50
23. 320; 280; 240; 200
24. 13 + y = 27, y = 14, 46 + 14 = 60 years old
25. 930 + y = 1,395, y = 465, 1,240 + 465 = 1,705 minutes
26. about 10,000 people or about 9,300 people

Chapter 3

Chapter 3
Lesson 1, page 32
1. 6   2. 12
3. 16   4. 20
5. 9   6. 6
7. 3 × 3   8. 5 × 4
9. 2 × 2

Chapter 3
Lesson 2, page 33
1. 42   2. 40
3. 72   4. 27
5. 64   6. 35
7. 49   8. 48
9. 45
10. 8 × 7 = 56 pizzas

Chapter 3
Lesson 3, page 34
1. 0   2. 8
3. 9   4. 0
5. 5   6. 5
7. 12   8. 0
9. 550   10. 26

Chapter 3
Lesson 4, page 35
1. 70   2. 55
3. 96   4. 99
5. 80   6. 121
7. 20   8. 77
9. 36   10. 72
11. 30   12. 60
13. Possible answer: I can break apart 12 into 10 + 2. Then I can multiply 10 × 8 and 2 × 8. That equals 80 and 16. Then I can add the products: 80 + 16 = 96.

Chapter 3
Lesson 5, page 36
1. Identity
2. Commutative
3. Commutative
4. Identity
5. Possible answer: He can arrange the desks in 7 rows of 4 each.

Chapter 3
Lesson 6, page 37
1. Associative
2. Distributive
3. Associative
4. Distributive
5. Possible answer: (10 × 5) + (2 × 5) = 60
6. Possible answer: (6 × 5) + (6 × 4) = 54

Chapter 3
Lesson 7, page 38
1. 3   2. 7
3. 5   4. 4
5. 6   6. 6
7. 3

Chapter 3
Lesson 8, page 39
1. 9   2. 7
3. 4   4. 9
5. 2   6. 6
7. Possible answer: 32 ÷ 4 would have a greater quotient because there are more being divided.

Chapter 3
Lesson 9, page 40
1. 12   2. 5
3. 2   4. 7
5. 3   6. 10

7. 60 ÷ 10 = 6 pies
8. 30 + 42 = 72; 72 ÷ 12 = 6 rows

Chapter 3
Lesson 10, page 41
1. 27; 36; 45
2. 60; 72; 84
3. 10; 8; 6
4. x ÷ 10
5. y ÷ 7
6. c × 2

Chapter 3
Lesson 11, page 42
Drawings will vary.
1. 3 × 6 = 18 books
2. 121 ÷ 11 = 11 display cases
3. x × 5 = 40, x = 8 rows

Chapter 3
Test, pages 43–44
1. 40   2. 45
3. 64   4. 33
5. 16   6. 36
7. 12   8. 21
9. 108   10. 8
11. 9   12. 12
13. 4   14. 3
15. 2   16. 9
17. 7   18. 5
19. $5.00
20. 24 ÷ 6 = 4 students
21. 7 × 5 = 35 buttons
22. Multiplication and division are inverse operations. Because 9 × 7 = 63, I know that 63 ÷ 7 = 9.
23. They are both correct. The Commutative Property says that factors can be multiplied in any order.
24. Identity
25. Distributive
26. Associative
27. Associative
28. Associative
29. Distributive
30. (12 × 6) + (4 × 6) = 72 + 24 = 96
31. 404
32. 10; 20; 30; 40; 50
33. 4; 10; 16; 22; 24
34. 11; 9; 6; 3; 1

Chapter 4

Chapter 4
Lesson 1, page 45
1. composite
2. prime

3. composite
4. composite
5. composite
6. prime
7. I disagree. A product of two prime numbers would have more than two factors. It would have at least 1, itself, and the two prime numbers you multiplied.
8. winter

Chapter 4
Lesson 2, page 46
1. 1 and 7
2. 1 and 6; 2 and 3
3. 1 and 16; 2 and 8; 4 and 4
4. 1 and 22; 2 and 11
5. 1 and 35; 5 and 7
6. 1 and 48; 2 and 24; 3 and 16; 4 and 12; 6 and 8
7. 1 and 100; 2 and 50; 4 and 25; 5 and 20; 10 and 10
8. Yes, he is correct. An even number of objects can always be arranged into 2 equal groups, so 2 will always be a factor for even numbers.

Chapter 4
Lesson 3, page 47
1. yes   2. yes
3. no   4. yes
5. no   6. yes
7. no   8. no
9. Possible answer: 71 is not a multiple of 2 because 71 is not an even number. All multiples of 2 are even numbers.
10. Possible answer: 45 is a multiple of 5 because the ones digit is a 5. All multiples of 5 have a 0 or a 5 in the ones place. 45 is also a multiple of 9 because 4 + 5 = 9. The digits of all multiples of 9 add up to 9 or a multiple of 9.

Chapter 4
Lesson 4, page 48
1. 2 × 3 = 6 cups of beans
2. 3 × 7 = 21 laps
3. 4 × 5 = 20 spoons
4. 11 × 9 = 99 shirts

Chapter 4
Lesson 5, page 49
1. 4 × 2 = 8 cups of peas
2. 1 × 13 = 13 feet long
3. 298 + y = 455; 455 − 298 = 157; y = 157; 475 + 157 = 632 yards
4. 5 × 9 = 45 marbles

Chapter 4
Lesson 6, page 50
1. y × 6 = 12; 12 ÷ 6 = 2 cups of peas
2. y × 8 = 32; 32 ÷ 8 = 4 blue horses
3. 36 ÷ 4 = 9 oranges in each bag

Chapter 4
Lesson 7, page 51
1. 3 × 4 = 12 lawns; 12 × 8 = 96 dollars
2. 6 × 5 = 30 flowers; 30 − 19 = 11 flowers are white

Chapter 4
Test, pages 52–53
1. composite
2. prime
3. composite
4. prime
5. composite
6. prime
7. 1 and 7
8. 1 and 18; 2 and 9; 3 and 6
9. 1 and 63; 3 and 21; 7 and 9
10. no   11. yes
12. yes   13. no
14. 2 × 5 = 10 cups of beans
15. 3 × 4 = 12 feet long
16. 56 markers
17. 44 years old
18. 30 pictures
19. y × 6 = 36; 36 ÷ 6 = 6 lions
20. 3 × 4 = 12 dogs; 12 × 9 = $108
21. 7 + 11 = 18 books; 18 ÷ 3 = 6 books on each shelf

Chapter 5

Chapter 5
Lesson 1, page 54
1. 52   2. 95
3. 132   4. 540
5. 130 chairs

Chapter 5
Lesson 2, page 55
1. 710
2. 813
3. 2,472
4. 2,058
5. 2,064
6. 7,902
7. 1,416

Chapter 5
Lesson 3, page 56
1. 5,160
2. 10,209
3. 14,370
4. 34,452
5. 29,876
6. 50,480
7. (2 × 2,367) + 1,983

Chapter 5
Lesson 4, page 57
1. 285   2. 374
3. 1,378   4. 154
5. 300   6. 144

Chapter 5
Lesson 5, page 58
1. 3,000
2. 3,600
3. 12,000
4. 56,000
5. 20,000
6. 15,000
7. Possible answer: When you multiply 4 by 5, the product is 20. It has a zero. So you have to add a zero to the product of 40 and 500.

Chapter 5
Lesson 6, page 59
1. 2,400
2. 2,500
3. 1,600
4. 2,400
5. 800
6. 1,200
7. 5 × 500 = 2,500 pounds of leaves

Chapter 5
Lesson 7, page 60
1. Sample estimate: $2,000; exact answer: $2,184; Sample estimate: $700; exact answer: $860
2. Sample estimate: 200; exact answer: 216 birds
3. Sample estimate: 18,000; exact answer: 17,517 children's books

Chapter 5
Test, pages 61–62
1. 64
2. 348
3. 776
4. 54
5. 315
6. 495
7. 1,596
8. 3,965
9. 5,000
10. 19,620
11. 63,920
12. 17,742
13. 3,192
14. 13,720
15. 36,092
16. 1,050
17. 377
18. 2,343
19. 527
20. 1,272
21. 686
22. 20,000
23. 18,000
24. 28,000
25. about 3,000
26. about 500
27. about 8,100
28. Sample estimate: 960; exact answer: 956 blankets; sample estimate: 556; exact answer: 554 blankets left to wash
29. Sample estimate: 160; exact answer: 192 cars
30. Sample estimate: 70; exact answer: 72 marbles; sample estimate: 130; exact answer: 137 marbles
31. Sample estimate: 300,000; exact answer: 303,587 people

Chapter 6

Chapter 6
Lesson 1, page 63
1. 14   2. 9
3. 10   4. 17
5. Drawings should show that he used 3 boxes.

Chapter 6
Lesson 2, page 64
1. 6R1   2. 6R3
3. 11R5   4. 7R5
5. 6R6   6. 40R1
7. 18R4   8. 9R5

Answers

9. Possible answer: When you divide 84 by 5, you get a greater quotient than when you divide by 9, because when you put fewer objects into each equal group, there will be more groups.

Chapter 6
Lesson 3, page 65
1. 37R1
2. 24
3. 40R2
4. 321 ÷ 3 = 107

Chapter 6
Lesson 4, page 66
1. 1,360R2
2. 569R5
3. 2,921
4. 1,362R1
5. $462

Chapter 6
Lesson 5, page 67
1. 2   2. 20
3. 200   4. 7
5. 70   6. 700
7. 60   8. 90
9. 40
10. 18 ÷ 6 = 3

Chapter 6
Lesson 6, page 68
Note: some estimates may vary.
1. about 40
2. about 26
3. about 110
4. about 25
5. about 250
6. about 90
7. 13; Possible answer: You need an exact answer. The question didn't ask "about how many hats." You need to know exactly how many hats. were sold.

Chapter 6
Lesson 7, page 69
1. 144 dolls were received; There is not enough information to answer this question. The number of shelves is not known.
2. This problem cannot be solved. The amount of money they start with is not known.

Chapter 6
Test, pages 70–71
1. 8   2. 7R4
3. 30R4   4. 2,137R1
5. 12   6. 22R1
7. 66R2   8. 993R3
9. 9   10. 14R5
11. 295R1   12. 1,562
13. 100   14. 60
15. 80
16. about 60
17. about 90
18. about 110
19. 30 sandwiches; He puts 7 sandwiches on each shelf. There are 2 sandwiches left over.
20. 43 comic books in each box with 4 left over
21. 1,350 people; 923 people did not buy popcorn
22. $x \div 6 = 25$; $x = 150$

Chapter 7

Chapter 7
Lesson 1, page 72
1. $\frac{1}{4}$ 2. $\frac{4}{6}$
3. $\frac{2}{3}$ 4. $\frac{1}{2}$
5. $\frac{1}{6}$ 6. $\frac{2}{8}$
7. Drawings should show 4 circles out of 10 shapes.

Chapter 7
Lesson 2, page 73
1. $\frac{2}{6}$
2. $\frac{1}{4}$
3. $\frac{2}{3}$
4. $\frac{8}{10}$
5. Possible answers: $\frac{1}{2}, \frac{5}{10}$

Chapter 7
Lesson 3, page 74
1. 15   2. 2
3. 6   4. 18
5. Possible answer: $\frac{6}{8}$
6. Possible answer: $\frac{3}{9}$
7. Possible answer: You can multiply 3 × 10 to get 30. That matches the number of students.

Then you can multiply 1 × 10 to get 10 and make an equivalent fraction, $\frac{10}{30}$. So there are 10 students whose names start with the letters M or C.

Chapter 7
Lesson 4, page 75
1. >   2. <
3. =   4. <
5. =   6. <
7. >   8. >
9. <
10. Possible answer: She needs to show the fractions on the same shape to compare them correctly.

Chapter 7
Lesson 5, page 76
1. <   2. >
3. >   4. >
5. =   6. <
7. More of the flag is colored gold than brown.

Chapter 7
Lesson 6, page 77
1. Possible answer: She is not correct. $\frac{1}{4}$ means the whole has been divided into 4 equal parts. The board shows 4 parts, but they are not equal parts.
2. Drawings should show parts of the same shape.
3. Possible answer: They are equivalent fractions, but they would not be equal if they were fractions showing different-sized wholes.

Chapter 7
Lesson 7, page 78
1. 12   2. 3
3. 8

Chapter 7
Test, pages 79–80
1. $\frac{2}{5}$ 2. $\frac{1}{4}$
3. Possible answer: $\frac{2}{8}$
4. Possible answer: $\frac{10}{12}$
5. Possible answer: $\frac{2}{3}$

6. Possible answer: $\frac{1}{2}$
7. <   8. =
9. >   10. <
11. Possible answer: He is correct. $\frac{3}{4}$ means the whole has been divided into 4 equal parts. The board shows 4 equal parts.
12. Drawings should show parts of the same sized shape.
13. 10   14. 6
15. $\frac{1}{5}$
16. $\frac{2}{6}$
17. $\frac{3}{4}$
18. $\frac{3}{6}$
19. >   20. >
21. =

Chapter 8

Chapter 8
Lesson 1, page 81
1. $\frac{4}{5}$
2. $\frac{2}{3}$
3. $\frac{3}{4}$
4. Possible answer: They are both correct. When you have 3 parts out of 3, you have the whole. 1 is $\frac{3}{3}$ in simplest form.

Chapter 8
Lesson 2, page 82
1. $\frac{1}{4}$
2. $\frac{2}{5}$
3. $\frac{1}{3}$
4. $\frac{1}{2}$
5. $\frac{3}{10}$
6. $\frac{1}{3}$
7. Possible answer: He can subtract numerators if denominators are the same. Since $1 = \frac{8}{8}$, he can subtract: $\frac{8}{8} - \frac{3}{8} = \frac{5}{8}$.

Chapter 8
Lesson 3, page 83
1. $\frac{2}{3}$
2. $\frac{2}{5}$
3. $\frac{4}{5}$
4. $\frac{3}{4}$
5. $\frac{3}{7}$
6. $\frac{7}{8}$
7. $\frac{4}{5}$
8. $\frac{8}{9}$
9. $\frac{2}{11}$
10. Possible answer: There is no more to decorate: $\frac{3}{7} + \frac{1}{7} = \frac{4}{7}$; $\frac{4}{7} + \frac{3}{7} = \frac{7}{7}$; $\frac{7}{7} = 1$. They have decorated the entire wall.

Chapter 8
Lesson 4, page 84
1. $\frac{1}{5} + \frac{1}{5} + \frac{1}{5} + \frac{1}{5}$
2. $1 + \frac{1}{3} + \frac{1}{3}$
3. $\frac{1}{3} + \frac{1}{3} + \frac{1}{3} + \frac{1}{3} + \frac{1}{3} + \frac{1}{3}$
4. $1 + \frac{2}{5} + \frac{1}{5}$; $1 + \frac{1}{5} + \frac{1}{5} + \frac{1}{5}$
5. $1 + 1 + \frac{3}{8} + \frac{2}{8} + \frac{2}{8}$
6. $1 + 1 + 1 + \frac{1}{2} + \frac{1}{2} + \frac{1}{2} + \frac{1}{2}$

Chapter 8
Lesson 5, page 85
1. 6
2. $7\frac{2}{3}$
3. $25\frac{6}{7}$
4. $31\frac{4}{5}$
5. $19\frac{3}{5}$ hours

Chapter 8
Lesson 6, page 86
1. $3\frac{5}{8}$
2. $4\frac{1}{2}$

3. $6\frac{1}{5}$
4. 49
5. Possible answer: $\frac{8}{8} = 1$; She can write 7 as $6\frac{8}{8}$ and then subtract. $6\frac{8}{8} - 2\frac{7}{8} = 4\frac{1}{8}$; $4\frac{1}{8}$ cups of water is left in the pitcher.

Chapter 8
Lesson 7, page 87
1. $3\frac{5}{8}$
2. $8\frac{1}{5}$
3. $59\frac{6}{7}$
4. $45\frac{1}{2}$
5. $15\frac{1}{4}$

Chapter 8
Lesson 8, page 88
1. $7\frac{7}{9}$ miles
2. $2\frac{1}{4}$ cups
3. $4\frac{1}{9}$ yards

Chapter 8
Lesson 9, page 89
1. $3 \times \frac{1}{4} = \frac{3}{4}$
2. $4 \times \frac{1}{5} = \frac{4}{5}$
3. $3 \times \frac{1}{6} = \frac{3}{6} = \frac{1}{2}$
4. $2 \times \frac{1}{2} = \frac{2}{2} = 1$
5. $5 \times \frac{1}{8} = \frac{5}{8}$
6. $4 \times \frac{1}{7} = \frac{4}{7}$

Chapter 8
Lesson 10, page 90
1. $\frac{8}{3} = 2\frac{2}{3}$
2. $1\frac{1}{5}$
3. $1\frac{1}{4}$
4. $1\frac{2}{7}$
5. $1\frac{3}{4}$
6. $1\frac{4}{5}$

Chapter 8
Lesson 11, page 91
1. $20 \times \frac{3}{4} = 15$ markers
2. $36 \times \frac{1}{3} = 12$ students

Chapter 8
Test, pages 92–93
1. $\frac{4}{5}$
2. $\frac{3}{4}$
3. $9\frac{1}{4}$
4. 72
5. $\frac{1}{3}$
6. $\frac{1}{2}$
7. $2\frac{1}{3}$
8. $11\frac{1}{5}$
9. $\frac{13}{16}$
10. $\frac{2}{7}$
11. $81\frac{7}{8}$
12. $45\frac{2}{3}$
13. $10\frac{7}{8}$ pounds
14. $1\frac{7}{9}$ cups
15. $\frac{1}{9} + \frac{1}{9} + \frac{1}{9} + \frac{1}{9}$
16. Possible answer: $\frac{5}{5} + \frac{1}{5} + \frac{1}{5} + \frac{1}{5}$
17. $1\frac{1}{3}$
18. $3\frac{3}{4}$
19. $1\frac{4}{5}$
20. $3\frac{1}{2}$
21. $17\frac{1}{8}$ miles
22. 19 trucks
23. $3\frac{1}{2}$ feet
24. $15\frac{1}{3}$ miles

Chapter 9

Chapter 9
Lesson 1, page 94
1. $\frac{20}{100}$
2. $\frac{40}{100}$
3. $\frac{10}{100}$
4. $\frac{5}{10}$
5. $\frac{4}{10}$
6. $\frac{1}{10}$

Chapter 9
Lesson 2, page 95
1. $\frac{6}{10}$
2. $\frac{70}{100}$
3. $\frac{8}{10}$
4. $\frac{5}{10}$
5. $\frac{68}{100}$
6. 1
7. Drawings will vary. Possible answer: The same amount of space is shaded in each square because $\frac{8}{10}$, $\frac{80}{100}$, and $\frac{4}{5}$ are equivalent fractions.

Chapter 9
Lesson 3, page 96
1. 0.40   2. 0.20
3. 2.90   4. 0.03
5. 3.71   6. 5.87

Chapter 9
Lesson 4, page 97
1. Point A on fifth tick mark on number line, begin count on first tick mark beyond zero.
2. Point B on third tick mark on number line, begin count on first tick mark beyond zero.
3. Point C on eighth tick mark on number line, begin count on first tick mark beyond zero.

Chapter 9
Lesson 5, page 98
1. 7 hundredths
2. 3 tenths

Answers

3. 4 hundredths
4. 0 hundredths
5. 4 ones
6. 3 tenths
7. 2 hundredths
8. 9 tenths
9. 3 ones
10. Answers will vary, but all should be 5-digit numbers with a 7 in the hundreds place, a 9 in the ones place, and a 2 in the hundredths place. Possible answer: 759.42

Chapter 9
Lesson 6, page 99
1. <
2. >
3. >
4. <
5. =
6. >
7. the blue jay

Chapter 9
Lesson 7, page 100
1. Possible answer: He is correct. Both Greg and George used miles to measure their distances. Since 3.48 is greater than 2.97, Greg hiked farther than George.
2. Possible answer: As a number, 0.85 is greater than 0.51, but Seina studied longer since a day is much longer than an hour.

Chapter 9
Lesson 8, page 101
1. $3.62
2. $7.48
3. $6.50
4. $8.03
5. Possible answer: You could use 5 dollars, 2 dimes, and 3 pennies.
6. Marcus

Chapter 9
Lesson 9, page 102
1. Draws new mark and label as shown in red.

- -|- - -|- - -|- - - - -
0   0.4   0.8

2. Draws new mark and label as shown in red.

- -|- - -|- - - - + - - -
0   0.2   0.5

3. Possible answer: It is not possible because there is not enough information. The exact number of teams is needed.

Chapter 9
Test, pages 103–104
1. $\frac{50}{100}$
2. $\frac{30}{100}$
3. $\frac{80}{100}$
4. $\frac{2}{10}$
5. $\frac{1}{10}$
6. $\frac{7}{10}$
7. $\frac{4}{10}$
8. $\frac{71}{100}$
9. $\frac{57}{100}$
10. Point A on sixth tick mark on number line, begin your count after the 0.
11. Point B on second tick mark on number line, begin your count after the 0.
12. Point C on fifth tick mark on number line, begin your count after the 0.
13. Draws new mark and label as shown in red.

- -|- - -|- - - - -|- -
0   0.4   1.2

14. 0.60
15. 2.78
16. 42.70 or 42.7
17. 5 hundredths
18. 4 tenths
19. 3 ones
20. >
21. =
22. <
23. $1.47
24. $2.05
25. Possible answer: I do not agree. 3.48 is greater than 2.39, but Benjamin's line is shorter because a foot is much shorter than a yard.

26. Yes, the comparison is valid, the weights of the two dogs are both expressed in pounds.

Chapter 10

Chapter 10
Lesson 1, page 105
1. about 1 centimeter
2. about 3 meters
3. about 16 centimeters
4. about 1 kilometer
5. 30,000 m
6. 500 cm
7. 8,000 m

Chapter 10
Lesson 2, page 106
1. Table should show the following data: Chase Street, 3 kilometers, 3,000 meters; Maple Street, 4 kilometers, 4,000 meters; Lower Bridge Street, 7 kilometers, 7,000 meters; Green Avenue, 8 kilometers, 8,000 meters
2. Table should show the following data: Height of Corn Plant, 2 meters, 200 centimeters; Height of Sunflower Plant, 3 meters, 300 centimeters; Height of Young Tree, 5 meters, 500 centimeters; Height of Old Tree, 18 meters, 1,800 centimeters

Chapter 10
Lesson 3, page 107
1. about 1 inch
2. about yard
3. about 10 feet
4. about 5 miles
5. 96
6. 21
7. 7,040
8. 468
9. 900
10. 31,680

Chapter 10
Lesson 4, page 108
1. Table should show the following data: Store to Park, 1 mile, 1,760 yd; Store to Home, 2 miles, 3,520 yd; Store to Library, 3 miles, 5,280 yd; Store to Bank, 4 miles, 7,040 yd
2. Second row in table: 3, 6, 9, 12, 15, 18; Third row in table: 36, 72, 108, 144, 180, 216

Chapter 10
Lesson 5, page 109
1. 48 square meters
2. 324 square centimeters
3. 128 square yards
4. 63 square meters

Chapter 10
Lesson 6, page 110
1. 40 meters
2. 46 kilometers
3. 32 inches
4. 20 in. of wood

Chapter 10
Lesson 7, page 111
1. $6\frac{7}{8}$ m
2. 208 m
3. $95\frac{2}{5}$ cm

Chapter 10
Lesson 8, page 112
1. $70\frac{4}{5}$ yd
2. $5\frac{1}{4}$ mi

Chapter 10
Lesson 9, page 113
1. 28 m
2. 48 sq ft
3. 112 sq km

Chapter 10
Test, pages 114–115
1. about 1 inch
2. about 2 meters
3. 9
4. 8,800
5. 144
6. 1,300
7. 48,000
8. 15,840
9. 396
10. 1,700
11. 26,400
12. Table should show the following data: Maple Tree, 32 feet, 384 inches; Oak Tree, 75 feet, 900 inches; Birch Tree, 56 feet, 672 inches; Pine Tree, 95 feet, 1,140 inches
13. Table should show the following data: Height of Garden Shed, 3 meters, 300 centimeters; Height

of Garage, 4 meters, 400 centimeters; Height of House, 22 meters, 2,200 centimeters; Height of Mailbox, 1 meter, 100 centimeters

14. 75 square inches
15. 169 square meters
16. 54 kilometers
17. $80\frac{5}{8}$ km
18. 58 yd
19. 88 sq m
20. 38 m
21. 25 feet

Chapter 11

Chapter 11
Lesson 1, page 116
1. about 1 gram
2. about 3 kilograms
3. 1,000
4. 4,000
5. 13,000
6. 57,000
7. 472,000
8. 397,000

Chapter 11
Lesson 2, page 117
1. about 1 pound
2. about 1 ounce
3. about 9 pounds
4. Second row in first table: 4,000, 6,000, 8,000, 10,000, 12,000 Second row in second table: 32, 48, 64, 80, 96

Chapter 11
Lesson 3, page 118
1. about 400 mililiters
2. about 200 mililiters
3. 2,000
4. 5,000
5. 13,000
6. 62,000
7. 113,000
8. 494,000

Chapter 11
Lesson 4, page 119
1. about 1 gallon
2. about 1 cup, or about 1 pint
3. 24
4. 12
5. 36
6. 272
7. 208
8. 5,840

Chapter 11
Lesson 5, page 120
1. about 3 months
2. Months: 24, 36, 48, 60, 72
Weeks: 104, 156, 208, 260, 312
Days: 730, 1,095, 1,460, 1,825, 2,190
Hours: 48, 72, 96, 120, 144
Seconds: 120, 180, 240, 300, 360

Chapter 11
Lesson 6, page 121
1. $32\frac{1}{2}$ kg
2. $33\frac{1}{4}$ kg
3. $31\frac{1}{4}$ kg
4. 1

Chapter 11
Lesson 7, page 122
1. $4\frac{1}{4}$ ft
2. 2 ft

Chapter 11
Lesson 8, page 123
1. $10\frac{3}{4}$ in
2. $9\frac{1}{4}$ in

Chapter 11
Test, pages 124 – 125
1. 3,000
2. 48
3. 106,000
4. 40
5. 10
6. 300
7. about 5 kilograms
8. about 10 tons
9. about 150 mililiters
10. 3 quarts
11. about 20 minutes
12. $11\frac{4}{5}$ L
13. $11\frac{2}{5}$ L
14. 1
15. 10 L
16. $1\frac{4}{5}$ L
17. $31\frac{3}{5}$ L
18. $5\frac{3}{4}$ cm
19. $2\frac{2}{4}$ cm or $2\frac{1}{2}$ cm

Chapter 12

Chapter 12
Lesson 1, page 126
1. A point is an exact location in space.
2. A line segment is part of a line. It has two endpoints.
3. A ray is a line with one endpoint. It goes on forever in the other direction.
4. C is point. $\overrightarrow{XY}$ is ray. $\overleftrightarrow{EF}$ is line. $\overline{JK}$ is line segment.

Chapter 12
Lesson 2, page 127
1. An angle is a figure formed by two rays that share the same endpoint.
2. A vertex is the endpoint shared by two rays.
3. angle, line, ray, angle
4.

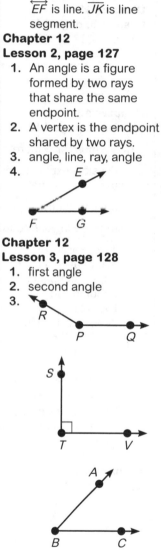

Chapter 12
Lesson 3, page 128
1. first angle
2. second angle
3.

Chapter 12
Lesson 4, page 129
1. check under second drawing
2. check under third drawing
3. check under first drawing

Chapter 12
Lesson 5, page 130
1. 2; no; no; trapezoid
2. 4; yes; yes; any or all of the following answers are correct: square, rectangle, rhombus, or parallelogram

Chapter 12
Lesson 6, page 131
1. equilateral and acute
2. isosceles and obtuse
3. scalene and acute
4. scalene and obtuse

Chapter 12
Lesson 7, page 132
1. right and isosceles
2. acute and isosceles
3. acute and equilateral
4. right and isosceles
5. right and isosceles
6. equilateral and acute
7. It is not possible for a triangle to have to right angles. All three angles in a triangle must add up to 180 degrees. Two right angles would leave the third angle as zero degrees. Two right angles would have parallel lines, not possible in a triangle.

Chapter 12
Lesson 8, page 133
1. 2
2. 0
3. 4
4. three lines of symmetry
5. one line of symmetry
6. no lines of symmetry

Chapter 12
Lesson 9, page 134
1. square, circle
2. 38
3. 24 pages

Chapter 12
Test, pages 135–136
1. A ray is a line with one endpoint. It goes on forever in the other direction.
2. An angle is a figure formed by two rays that share the same endpoint.
3. the middle drawing
4. the lines in the left-hand drawing are parallel lines
5. 4; no; no; parallelogram

Answers

6. 4; no; yes; parallelogram
7. isosceles and right
8. equilateral and acute
9. obtuse and scalene
10. Possible drawing:

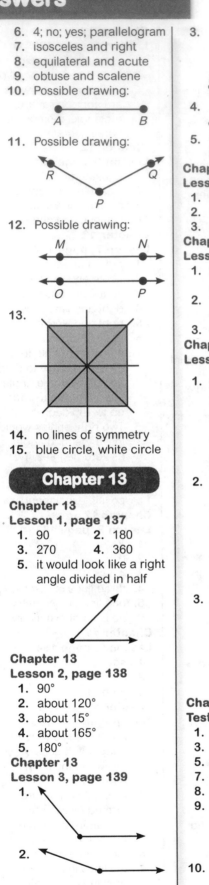

A   B

11. Possible drawing:

R   Q
P

12. Possible drawing:

M   N
O   P

13.

14. no lines of symmetry
15. blue circle, white circle

Chapter 13

Chapter 13
Lesson 1, page 137
1. 90   2. 180
3. 270   4. 360
5. it would look like a right angle divided in half

Chapter 13
Lesson 2, page 138
1. 90°
2. about 120°
3. about 15°
4. about 165°
5. 180°

Chapter 13
Lesson 3, page 139
1.
2.

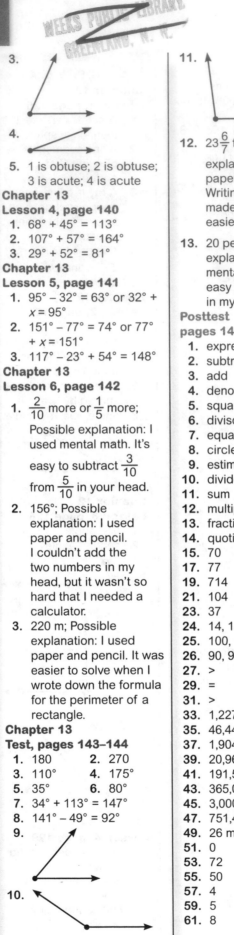

3.
4.
5. 1 is obtuse; 2 is obtuse; 3 is acute; 4 is acute

Chapter 13
Lesson 4, page 140
1. 68° + 45° = 113°
2. 107° + 57° = 164°
3. 29° + 52° = 81°

Chapter 13
Lesson 5, page 141
1. 95° − 32° = 63° or 32° + x = 95°
2. 151° − 77° = 74° or 77° + x = 151°
3. 117° − 23° + 54° = 148°

Chapter 13
Lesson 6, page 142
1. $\frac{2}{10}$ more or $\frac{1}{5}$ more; Possible explanation: I used mental math. It's easy to subtract $\frac{3}{10}$ from $\frac{5}{10}$ in your head.
2. 156°; Possible explanation: I used paper and pencil. I couldn't add the two numbers in my head, but it wasn't so hard that I needed a calculator.
3. 220 m; Possible explanation: I used paper and pencil. It was easier to solve when I wrote down the formula for the perimeter of a rectangle.

Chapter 13
Test, pages 143–144
1. 180   2. 270
3. 110°   4. 175°
5. 35°   6. 80°
7. 34° + 113° = 147°
8. 141° − 49° = 92°
9.
10.

11.

12. $23\frac{6}{7}$ ft; Possible explanation: I used paper and pencil. Writing an equation made the problem easier to solve.
13. 20 people; Possible explanation: I used mental math. It was easy to divide 140 by 7 in my head.

Posttest
pages 145–150
1. expression
2. subtract
3. add
4. denominator
5. square
6. divisor
7. equal groups
8. circle
9. estimate
10. divide
11. sum
12. multiply
13. fraction
14. quotient
15. 70   16. 51
17. 77   18. 19
19. 714   20. 808
21. 104   22. 43
23. 37
24. 14, 16, 18
25. 100, 110, 120
26. 90, 95, 100
27. >   28. <
29. =   30. <
31. >   32. <
33. 1,227   34. 50,291
35. 46,446   36. 20,931
37. 1,904   38. 130,426
39. 20,969   40. 1,935
41. 191,585   42. 758,253
43. 365,046   44. 292
45. 3,000   46. 589,000
47. 751,400   48. 25,200
49. 26 m   50. 35 sq ft
51. 0   52. 16
53. 72   54. 48
55. 50   56. 5
57. 4   58. 8
59. 5   60. 11
61. 8   62. 6

63. Distributive Property of Multiplication
64. composite
65. prime
66. prime
67. composite
68. 1 and 12, 2 and 6, 3 and 4
69. 1 and 19
70. 2 × 4 = 8 cups of peas
71. y × 6 = 54; 54 ÷ 6 = 9 pickles
72. Possible answer: $\frac{6}{8}$
73. Possible answer: $\frac{3}{5}$
74. Possible answer: $\frac{1}{9}$
75. $\frac{3}{5}$
76. $\frac{4}{5}$
77. $31\frac{4}{15}$
78. 0.50
79. 0.30
80. 0.78
81. 0.8
82–84:

A   C   B
1   2

85. 300 cm
86. 744 in.
87. 11,000 g
88.
89.
90.
91.

92. about 40°
93. about 155°
94. 80°
95. 88°